Visual Basic Telephony

Using Off-the-Shelf Components to
Build Windows-Based Telephony Applications

Chris Brookins
Krisztina Holly

Stylus Innovation, Inc.

Preface

Four years ago, three MIT graduate students invented a new technology enabling people to shop from home with a swipe of a barcode wand. The product was a pen-sized barcode-to-touchtone conversion device called The Stylus. Users would scan barcodes out of a catalog that listed groceries, office supplies, etc. The barcodes would be converted into a series of touch tones, sent over a phone line, and decoded by a telephony application running on the store's computer. The computer would look up information in a database, such as the price of the selected item, and speak that information to the customer. For example, after scanning a barcode next to a catalog listing for soup, the user might hear, "Campbell's Chicken Noodle Soup, 10 ounces, eighty-nine cents.". The three students, John Barrus, Mike Cassidy, and Krisztina Holly, applied for a patent, and started Stylus Innovation with a $500 investment each.

After successfully licensing the wand, Stylus hired Chris Brookins to develop the telephony application that decoded the touch-tones sent by the device. In the process of building this software and learning about computer-phone integration, Chris used several then state-of-the-art telephony development tools. Unfortunately, he found them awkward and inadequate for development, especially when compared to the GUI client/server tools with which he was familiar.

The telephony tools available in 1993 could not easily interact with mainstream databases or other applications such as report writers. In addition they either had significant runtime fees, used unusual proprietary languages, or were fill-in-the-blank application generators with little flexibility. Finally, they all ran in DOS, UNIX, or OS/2, which prevented them from working with most Windows applications.

The client/server industry was revolutionized when Visual Basic was introduced in 1991. With its industry-standard BASIC language, open architecture for add on components, incredible data access, and no runtime fees, Visual Basic quickly became the most popular Windows development environment. Chris felt that if Stylus could create a telephony component enabling Visual Basic to be a powerful telephony development tool, Stylus would have an equally dramatic effect on the telephony tools industry.

At that point we rolled up our shirtsleeves and developed Visual Voice, the first commercial toolkit that added the power of telephony to Visual Basic. Stylus launched Visual Voice at Fall Comdex '93 and was instantly deluged by customers and the press. Since then, Visual Voice has been adopted by thousands of customers and won numerous awards, including *Byte's* "Best of Fall COMDEX '93 Finalist," *Windows Magazine's* "Win 100 '95," *Teleconnect's* "Editor's Choice," *Computer Telephony's* "Product of the Year, 1994," and *Byte's* "1994 Award Of Excellence" honoring the 10 most significant technologies in the PC industry.

When we were approached to write this book, under an extremely tight deadline, we did not think it was really possible -- but, once again, the Stylus team pulled together. This book attempts to cover a wide range of topics in both the Windows and telephony areas. While we would have liked to go into greater depth in many topics, this book should serve as a good overview of Windows, Visual Basic, and telephony development.

We hope that all readers will feel free to notify us or the publisher of any inaccuracies, positive or negative feedback, and suggestions for improvement in future revisions of the book. You can write:

Stylus Innovation, Inc.
Attn: Visual Basic Telephony book
One Kendall Square, Building 300
Cambridge, MA 02139
phone 617.621.9545
fax 617.621.7862

Most importantly, the creation of this book was absolutely a group effort. First an incredible thanks to the entire Stylus team, especially Nancy Baron, Mike Cassidy, and Julie Vincent for researching and writing several key sections. Also, thanks to Harry Newton for offering us the opportunity to write this book, and Bob Edgar for providing us our initial telephony education through his wonderful book, *PC-Based Voice Processing*. Thanks also to Dialogic Corporation for guiding us during the early days of Visual Voice. And, of course, a big thank you to all our friends and family that gave us support during the creation of this book.

Chris Brookins
Krisztina Holly

Stylus Innovation, Inc.

Contents

INTRODUCTION TO COMPUTER TELEPHONY

WINDOWS AND TELEPHONY

VISUAL BASIC

TELEPHONY CONCEPTS

DEVELOPING TELEPHONY APPLICATIONS

VISUAL VOICE

TAPI IN DEPTH

TELEPHONY HARDWARE

CHAPTER ONE

Introduction to Computer Telephony

1.1 What is Computer Telephony?

Computer telephony (pronounced "teh - **leh**' - fuh - nee") allows people to interact with a computer through their telephone. Computer telephony systems are everywhere; you probably come into contact with them several times each day. When you call someone and enter the extension using touch-tones, you are using a voice mail/auto attendant application. When you call a phone number to hear the weather forecast, you are using an audiotex application. When your phone rings while you are eating dinner and someone tries to sell you insurance, the call was probably placed by a predictive dialing system.

The installed base of telephony applications is steadily increasing. Auto attendants used to be limited to large or mid-sized firms. Today, however, many companies with fewer than a dozen employees are installing voice mail/auto attendant systems. IVR systems used to be limited to touchtone banking at large banks. Today, movie theaters are commonly using audiotex systems that let you spell the name of a movie and hear show times, and even local golf ranges are installing IVR tee-time reservation systems. Fax-on-demand systems used to be restricted to technical support questions or brochures at large corporations. But today, even local realtors are building systems that will automatically fax you descriptions of properties that meet your criteria.

A typical computer telephony system performs three functions:

- Process an automated phone conversation with a caller
- Monitor and log calls
- Access external data or processing

A computer telephony application communicates with callers through various functions, such as:

- *Picking up the line:* to answer an inbound call
- *Placing an outbound call:* to initiate a telemarketing call, to remind patients about appointments, *etc.*
- *Playing voice files:* to report event information, quote stock prices, *etc.*
- *Prompting for digits:* to gather an order number, get the caller's PIN code, *etc.*
- *Prompting for a message:* to record a voice mail message, record special order instructions, *etc.*
- *Handling call flow through menus:* to determine which service or selection the caller wants
- *Detecting a hangup:* to determine if the caller terminated the call

1.2 How Do I Get Started?

Computer telephony is a powerful trend in communications, and new applications are being developed and installed everywhere. To implement your own computer telephony system, you have four main choices:

- Buy a turnkey application
- Use a fill-in-the blanks application generator
- Use a high-level development toolkit
- Control the voice board directly

Turnkey applications are easy to use, but they might not always suit your needs in terms of capacity, functionality, or flexibility. Fill-in-the-blank application generators allow you to simply define your application, but still limit you to the options they present. If an option you need is not available, you may be stuck.

For maximum flexibility, you can control the voice board directly. To do this, you must interface to the voice board driver, which is very low-level and difficult to work with. Consequently, it is helpful to have a higher level language or toolkit which you can use to develop your applications.

However, until very recently, you had to be an expert to create a computer telephony system with most toolkits. But today, it is relatively easy for even a novice to develop sophisticated telephony applications using off-the-shelf components.

This book shows how to use the ease, power, and flexibility of Visual Basic to create your own Windows-based telephony applications, whether they be simple, single-line voice mail applications or mission-critical call distribution systems. It is organized in the following Chapters:

Windows and Telephony explains Microsoft Windows, its connectivity options, its tools, and why it makes a great telephony platform.

Visual Basic describes the tool's environment, language, data access, third party options, and how to interface to telephony,

Telephony Concepts covers how the phone system works and the describes the technologies used when building systems.

Developing Telephony Applications describes common applications and the necessary components and techniques used to develop them.

Visual Voice provides an overview of the telephony toolkit for Visual Basic, Visual Voice.

TAPI in Depth covers the standard method for performing Windows telephony in greater detail.

Telephony Hardware describes the type of hardware available and how to get more information on the vendors options.

CHAPTER TWO

Windows and Telephony

2.1 Why Windows?

Microsoft Windows has undoubtedly become the most popular operating platform ever developed. Its easy-to-use graphical interface dominates both office and home desktops. Still, many people wonder, "Why do I need Windows for my telephony platform? Telephony applications have audio, not graphical, interfaces!"

Up until the end of 1993, DOS and OS/2 predominated the voice processing industry. This cycle was in no doubt fueled by voice board manufacturers who had no Windows drivers and tool makers that were entrenched in a DOS state-machine world. Recently, however, Windows-based telephony applications have flooded the market.

Several factors make Windows an ideal platform for developing and running telephony applications:

- Unmatched connectivity options
- Huge installed base and popularity
- Many highly productive development tools
- Multi-tasking to handle multiple phone calls simultaneously
- Graphical user interface for professional looking applications

This chapter will look at each of these factors in more detail.

2.2   Connectivity Options

Because Windows is the most popular operating environment, it stands to reason that third-party developers spend more time creating products for Windows than for other platforms. This is a significant reason for choosing Windows as a telephony operating platform.

Telephony and fax applications require easy access to business data. Windows provides more database connections, network protocol access, and mid-range and host system connectivity that any other platform. Companies like Oracle, Sybase, Tandem, Informix, DCA, and even IBM concentrate the majority of their client software development on Windows.

2.2.1   Windows Plumbing

The modular architecture of Windows allows you to use pre-built components to easily develop and deploy applications that can work together. Windows provides the plumbing to make modular connectivity options accessible to development tools:

- *Dynamic Link Libraries (DLLs):* Sharable libraries, typically created in C, that are dynamically loaded only when the DLL is referenced. If the DLL is already loaded and being used by another application, it is shared.

- *Dynamic Data Exchange (DDE):* A well-defined communications protocol between concurrently running Windows applications. Telephony applications can use DDE to request information from a database, data from a host emulation package, and more.

- *Object Linking and Embedding (OLE):* A standard way for applications to imbed the functionality of other applications inside of them. With OLE, an application can control and imbed any portion of any other OLE application through OLE automation.

In addition, Microsoft is also beginning to distribute more connectivity options with the Windows for Workgroups operating platform itself. For example, Windows At Work Fax, included with Windows for Workgroups, enables your client applications to use the Microsoft Messaging API to send faxes from across the network (See 2.3.2 below and 5.5.1 for more information).

2.3 WOSA

In order to foster a heterogeneous programming environment on its Windows operating systems, Microsoft developed *WOSA*, the Windows Open System Architecture. WOSA sets common standards for applications to interface to corporate data and other resources. Currently, WOSA includes common interfaces for databases (ODBC), messaging services (MAPI), telephony services (TAPI), real-time financial data, and more.

To take advantage of WOSA's true inter-changability of parts, both the application and the resource (database, mail server, voice board, etc.) must be coded to support the given WOSA service. Therefore, the WOSA model encompasses both the applications side (application programming interface, or *API*) and the resource side (service provider interface, or *SPI*). Windows system software provides the glue between the API and SPI implementations.

Whether you are personally a Microsoft friend or foe, there is no denying that Microsoft's standards have had major influence in the industry. Thus, while some standards have to struggle to gain acceptance, WOSA has third-party vendors rushing to support it.

2.3.1 ODBC

The Open Database Connectivity (ODBC) standard is meant to provide a standard interface across all database and file management systems. As a consequence, telephony developers can use the same ODBC interface in their applications to access any database or file management system that supports the ODBC standard. ODBC is widely supported in the industry; an ODBC application can retrieve and update data from practically any database in the industry, including Access, Oracle, DB2, Rdb, and many more.

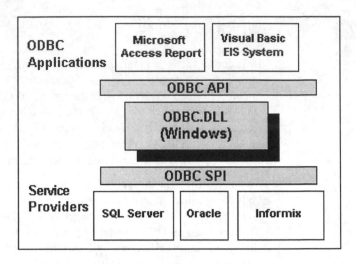

2.3.2   MAPI

Microsoft's *MAPI* stands for "Messaging API." Windows telephony applications can integrate MAPI into their applications for a variety of reasons, including sending faxes using Microsoft's Windows at Work Fax Gateway and interfacing to e-mail packages that support MAPI. Most popular Windows mail packages now support MAPI.

An increasingly popular concept is that of *unified messaging* (See Section 5.1.5 for more details). Unified messaging strives to provide a single interface for users to send and receive voice mail, faxes, and e-mail. The MAPI interface enables a telephony developer to tie into multiple messaging sources to create a unified messaging application. For example, a Windows telephony application can use the MAPI interface to embed a voice mail message or fax file into an e-mail note.

2.3.3   TAPI

Microsoft and Intel jointly developed the Telephony API (TAPI) to provide a standard interface across telephony hardware platforms. A TAPI application will be able to run unmodified on any TAPI-compliant voice board, PBX, or other device. For complete information on the Telephony API, see Chapter 7.

2.3.4   Other WOSA Standards

WOSA also specifies connectivity APIs for real time financial data, and SNA host connectivity.

2.4   Windows Development Tools

In addition to providing their own tools for Windows development, Microsoft has attracted the vast majority of third-party development tools vendors to the Windows platform. Windows programmers can find a wide range of tools, including graphical report writers, database development tools, client/server development tools, and object-oriented platforms. The following table lists some popular Windows development tools.

Desktop Databases	Client/Server Tools	Object-BasedTools
Microsoft Access	Microsoft Visual Basic	Microsoft Visual C++
Microsoft Foxpro	Powersoft PowerBuilder	Borland C++
Borland dBase for Windows	Gupta SQL/Windows	Symantec C++
Borland Paradox for Windows	Borland Delphi	Park Place SmallTalk

The competition in the Windows tools market is very fierce. Consequently, tool manufacturers have packed their products full of the latest features in the hopes of attracting development dollars. Although Windows development tools range from non-proprietary C++ and BASIC to very graphical proprietary systems, they all have many things in common:

- Graphical user interface design tools
- Event-driven architecture
- Many connectivity options (minimally ODBC support)
- On-line help
- No runtime fee policy

2.4.1   Visual Basic

Of the numerous development tools available, Visual Basic is the most popular. Over 1 million developers worldwide use Visual Basic to develop Windows applications. Visual Basic is based on the non-proprietary BASIC language. Because of the hundreds of developers that have created add-on custom controls, Visual Basic provides unparalleled connectivity options (even compared to other popular Windows development tools).

While all of the Windows development tools would be appropriate for developing telephony applications, the remaining chapters of this book will concentrate on telephony and how it applies to Visual Basic. For more information specifically about Visual Basic, read Chapter 3.

2.4.2   Componentware

An emerging trend in Windows platforms is the concept of function-specific components. Visual Basic was the first to introduce its support for *custom controls* (also known as VBXs). Visual Basic was designed in such a way that additional functionality could be "snapped-in" to the development environment by the addition of these Visual Basic controls.

Once installed, add-on custom controls are identical to controls shipped with Visual Basic. (For more information about custom controls and how they integrate into Visual Basic, refer to Section 3.2.2.) A custom control appears in the Visual Basic toolbox like the other tools, includes events that the program can respond to, provides properties that change the object's characteristics, and integrates its on-line help with that of Visual Basic. Examples of Visual Basic custom controls include spread-sheet objects, database connectivity modules, and telephony controls.

Microsoft has now extended the concept of componentware with the introduction of *OLE Controls* (or OCXs). OLE Controls are based on Microsoft's Object Linking and Embedding (OLE) architecture.

In the past, non-Microsoft development tools that wanted to support VBX's (e.g. PowerBuilder) had to implement their support without help from Microsoft. OLE, however, is a published specification. Microsoft is actively encouraging tools and applications vendors to be OLE servers, supporting the snap-in capability of OLE Controls.

The OLE control model has three distinct advantages over the VBX model:

- *Development tool portability:* An OLE control will be able to be plugged-in to a wide variety of tools (e.g., Access, FoxPro, Lotus Notes, Borland Delphi, etc.) in addition to Visual Basic. Microsoft, Lotus, and Borland (to name a few) are all planning support for OLE controls.

- *Uniformity across development tools:* An OLE control's visual interface is the same no matter in which environement it is used. Instead of requiring all OLE server developers to create their own interfaces for setting control properties, the OLE standard has predefined "property pages" that move with the OLE control across development environments. Consequently, you will see the same screens whether you are using Visual Voice in Excel or PowerBuilder.

- *Pathway to Microsoft 32-bit operating platforms:* Although VBXs will run on Windows NT and Windows '95, they will always remain 16-bit. 32-bit OLE Controls will take advantage of the pre-emptive multitasking capabilities that the newest operating systems provide.

2.5   Multitasking

The current versions of Windows provide a multi-tasking layer above DOS. Windows multi-tasking lends itself to telephony applications for several reasons:

- You can run multiple voice processing applications at once in Windows. For example, you can run voice mail on lines 1 - 4, and an order entry application on lines 5 - 10.

- Your telephony application can interface to other running Windows and DOS applications on the same PC. For example, a real-estate fax-on-demand system may ask a caller to enter their desired house-buying criteria. The fax-on-demand application can pass the data to a report writer running on the same system to generate a formatted report of available homes.

- You can usually run other applications on your telephony server PC without having to bring-down the telephony application.

Although Windows is a multi-tasking environment, it does not provide true pre-emptive multi-tasking capabilities that are available on Windows NT, OS/2, and UNIX. Windows does not automatically guard against one application gobbling up processor time or corrupting the memory addresses of another application. As a result, all running applications must be well-behaved for Windows systems to stay responsive. Well-behaved applications are designed to give the processor time to multi-task and stay within their own address spaces. Any other applications you plan to run alongside your telephony applications should be tested in advance.

2.6   Graphical User Interface

Graphical user interfaces (or *GUIs*) and their paradigms of check boxes, radio buttons, and list boxes have become the standard for computer application users today. Although graphical user interfaces are of no consequence to the callers of a telephony application, they can make the application much easier to administer and market.

One of the issues with proprietary voice mail systems is the difficulty of system administration. Typically, configuring a PBX or key system requires pulling out the manual and entering bizarre sequences of touch-tones recognizable only to a particular vendor's PBX.

By using a Windows development tool, you can build telephony applications like voice mail that have easy-to-use GUIs to the phone system. Your GUI can include a list of users, can interface to a database to automatically bring up a list of available extensions, and can offer integrated on-line help. You can even incorporate drag-and-drop features to assign users to voice mail groups or departments.

GUI tools also enable you to design and develop very professional-looking applications. Although appearance may be of secondary importance to ease-of-use for some in-house applications, a good presentation is essential for resale applications. An easy-to-use interface with appealing graphical elements and controls attract a customer's attention. A well-conceived and executed user interface portrays a solid product. Today's buyers want systems that regular PC users can easily update and control, with minimal training.

2.7 The Future of Windows

In time, Microsoft 32-bit operating platforms will replace Windows 3.1 as the dominant platform. Microsoft is targeting Windows NT for servers and high-powered users while targeting Windows '95 for the desktop. Windows NT is already commanding a significant slice of the server operating system market. Given Microsoft's track record, Windows '95 will be a tremendous seller starting the day it is released.

The pre-emptive multi-tasking abilities that Windows NT and Windows '95 include are a boon for telephony applications. These operating systems will be able to handle more simultaneous phone lines and other processes than Windows 3.1.

In order for Windows NT and Windows '95 to take off as telephony platforms, telephony development tools and hardware platforms must be available. After some delay, the major board manufacturers are coming out with Windows NT drivers, with Windows '95 drivers soon to follow. Dialogic and Rhetorex already have drivers for Windows NT. TAPI drivers are just starting to emerge for Windows 3.x, and by the time Windows '95 ships, hundreds of options should be available.

Fortunately, most Windows development tools today will readily operate on Microsoft's 32-bit platforms. PowerBuilder already works under Windows NT; Microsoft has stated that the next versions of Visual Basic, FoxPro, and Visual C++ will be available on both 32-bit platforms. Applications written today for Windows 3.1 should be easily portable to the new 32-bit platforms.

CHAPTER THREE

Visual Basic

3.1   Overview

Microsoft introduced Visual Basic in 1991 as a general purpose development environment for quickly creating Windows applications. Soon afterwards, it became the most widely used graphical development tool of all time. Currently in version 3.0, Visual Basic has an installed base of over 1.5 million users.

Visual Basic's popularity stems from several factors:

- Easy to use environment
- Standard language
- Excellent data access
- Completely open mechanism for third-party extensions (i.e., *custom controls*)

With a standard way of extending the Visual Basic environment, hundreds of third parties have flocked to support Visual Basic with a myriad of add-ons, effectively removing any "walls" within the Visual Basic environment.

Applications developed with Visual Basic are compiled into a Windows EXE that you can distribute with Visual Basic's runtime DLL at no additional charge. Visual Basic is today viewed as the ubiquitous, industry standard, non-proprietary way to develop Windows applications. For these exact reasons, Visual Basic is also viewed as an excellent platform for developing computer telephony applications.

3.1.1   Using Visual Basic

Developing a Visual Basic application is a *visual* process, as easy as painting a *form,* or window, on your screen. You simply drop the appropriate buttons, listboxes, and other *controls* onto the form to customize the application. Finally, you insert BASIC code to define how your application behaves.

Forms and controls are *objects*, each with their own *properties*, *events*, and *methods* that define and control their behavior. Properties are settings like variables (e.g., button size and caption). Events are special subroutines that are executed every time a special condition is met (e.g., a mouse click). Methods are special actions that objects can perform that you can call (e.g., the show method for forms, the refresh method for databases).

Visual Basic's object-based structure enables you to point-and-click to quickly create and customize a GUI and associated connectivity.

Here is a 30,000 foot view of how you would typically develop a Visual Basic application:

1. Create one or more Forms (Windows).

2. Add the necessary controls to the form and customize them.

3. Insert the appropriate code into the event subroutines to configure the desired behavior of your application.

4. Run the application from within Visual Basic to debug it.

5. Compile the application to an EXE, test, and distribute it.

3.2   The Visual Basic Environment

Visual Basic includes a highly graphical environment to make it easy and fun to create sophisticated applications. The following screenshot shows the Visual Basic development environment:

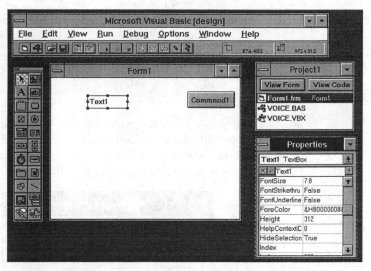

At the most basic level, you control your development session from Visual Basic's toolbar shown at the top of the screen. From here, you can press the VCR play, stop, and pause buttons to control the running of your application. You can also view any of the other Visual Basic windows that aid in the development process, including:

- Project window
- Toolbox
- Code window
- Property window

3.2.2   Project Window

You can view a list of your application's components in the Visual Basic's Project window.

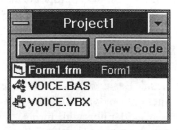

You can add, edit, or remove any component from this window. The components include:

- Forms (.FRM)
- Modules (.BAS)
- Custom Controls (.VBX)

Forms

Forms ▯ are graphical windows you can build with traditional GUI controls or custom controls. Forms contain associated code describing how the window and controls behave. The example form below has been designed to look like a phone keypad.

The form above includes several controls, including command buttons, labels, and 3D panels. This form's Caption Property has been set to the string "Virtual Phone." Other properties describe the form's size, position, and color. Every form has associated Visual Basic code that describes the behavior of the form and related controls. To see the corresponding code, simply double click on any control or on the form itself, and the Code window appears with the appropriate subroutine loaded. The Code window is described in further detail in Section 3.2.5 below.

Modules

Visual Basic *modules* ▧ assemble any additional subroutines, functions, or variables that are not related to a specific form or control. Unlike with forms, you can make any code or variables placed within a module *global*, so they are accessible by other forms and modules.

Controls

Visual Basic includes a set of standard controls that appear in your toolbox (described in the next section below). This toolbox can be extended with *custom controls* ⚘ , special libraries that add additional functionality to your Visual Basic applications. Several custom controls are included with Visual Basic, and hundreds of others can be inexpensively purchased from third-party vendors. In the Virtual Phone form above, the command buttons and labels are standard controls, while the embossed areas are instances of the 3D Panel custom control included with the Professional Edition of Visual Basic.

While many custom controls have a specific graphical look (such as a button or a listbox), some stay hidden at runtime. The last control on the Virtual Phone, labeled "TEST," is a custom control provided with Visual Voice (described in Chapter 6) that does not display when the application is running but simply provides special functions like background communications.

3.2.3 Toolbox

Visual Basic's *toolbox* displays the controls you can add to your forms. Visual Basic comes in two editions: Standard and Professional.

Standard Edition

The toolbox on the next page contains all of the built-in controls that the Visual Basic Standard Edition provides:

19

		Picture Box
Label		Text Box
Frame		Command Button
Check Box		Option Button
Combo Box		List Box
Horizontal Scroll Bar		Vertical Scroll Bar
Timer		Drive List Box
Directory List Box		File List Box
Shape		Line
Image		Data Control
Grid		OLE 2.0

Professional Edition

The Professional version adds the following capabilities to the Standard Edition:

- Additional custom controls
- Enhanced database support
- Control Development Kit (CDK) that you can use to build your own custom controls in C
- The Windows Help Compiler to build your own on-line help
- Crystal Reports for generating and printing reports from a database

After you have added all of the Professional Edition controls to your project, your toolbox looks like the toolbox on the next page.

Common Dialog

Graph

Serial Comms

Masked Edit

OLE Client

3D Frame

3D Panel

Animated Button

Crystal Reports, Gauge

Grid, MCI Control

MAPI Session, MAPI Mess

OLE 2.0, Outline Control

Spin Button, 3D Check Box

3D Option Button, 3D Com

3D Group Push Button

Third-Party Controls

Visual Basic is such a compelling development platform because you can extend the toolbox with controls purchased from over 300 third-party Visual Basic add-on vendors. Once you add a third-party custom control, it appears in the toolbox along with the other controls.

Just like the controls included with Visual Basic, these controls expose their own properties and events. Since the interface for third-party controls and Visual Basic's built-in controls are the same, you can immediately incorporate the technology offered by a third-party and feel confident that it will be easy to learn and use.

3.2.4   Property Window

To change a property of an object at design time, simply select the object and change the appropriate value in the Visual Basic Properties window. The top of the window displays the selected object's name and type; beneath the name appears a list all of the applicable properties.

Depending on the property, you can change its value in one of three ways: typing in a value, picking a value from a drop down list of valid values, or double-clicking to display an associated dialog (e.g. a color palette).

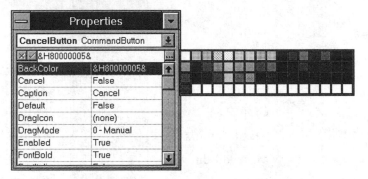

If several objects have been selected, their common properties are displayed. Changing a property will effect all of the objects. This allows you to quickly change the color, font, or alignment across a group of related controls in one easy step.

3.2.5   Code Window

The Code window displays the logic that controls your application behavior; you should write or paste all of your code into this window.

Every form and module has its own associated Code window that can contain program declarations or procedures.

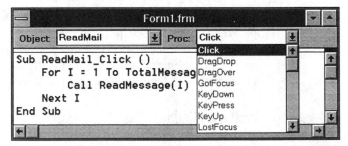

The code visible in the above example will be executed whenever the user clicks the ReadMail button (i.e., the control's Click Event is triggered). Each event subroutine for each object can have its own code.

Using the drop down lists at the top of the window, you can view all of the application's associated code, including local subroutines, variable declarations, and each object's events.

Visual Basic's user interface makes coding easy. The code window automatically color-codes anything you type, making the logic easy to read and debug. You can jump from a procedure reference to its declaration with a simple key press.

Visual Basic incrementally compiles any changes you make to your code, which prevents time-consuming recompilation of your entire application every time you run it.

There are four ways to make something occur in a Visual Basic application at runtime:

- Calling a method
- Initiating an action
- Setting a property
- Calling a procedure

Calling Methods

Forms and most controls have *methods* -- built-in procedures associated with that particular control. For example, the following code uses the Show method to display the window called Form1:

```
Form1.Show
```

The Hide method performs the opposite function. Methods vary from object to object.

Initiating Actions

Because Microsoft's Visual Basic custom control specification does not allow 3rd party controls to have their own methods, they often expose a property named Action to implement the same functionality. To make the custom control perform various functions, simply set the Action Property to the corresponding constant.

For example, the Common Dialog control (CMDialog) has an Action Property that, when set to any valid value, causes a standard Windows dialog box to appear. The following code will display a standard Windows File Open dialog box so a user can select and open a file:

```
Cmdialog1.Action = 1
```

On the other hand, setting the Action Property to 3 will show a standard Windows Color dialog box.

Note that the Microsoft OLE Control specification allows methods, so OLE Control versions of custom controls should not need Action Properties in the future, though they will probably be retained for backward compatibility.

Setting Properties

Properties define an object's characteristics and can be referred to like regular variables. For example, the following line of code moves the window Form1 towards the bottom of the screen by 100 units:

```
Form1.Top = Form1.Top + 100
```

As shown above, some properties have an immediate effect when set at runtime. In contrast, some properties do not have an immediate effect, but instead serve as inputs to a subsequent action. Once the action is called, the setting takes effect. For example:

```
Voice1.PhoneNumber = 5551212
Voice1.Action = VV_CALL
```

Calling Procedures

You can build your own functions and subroutines and call them from your application. These procedures, of course, can use methods, actions, and/or properties to control execution within them.

3.3 The Visual Basic Language

Four distinct features of Visual Basic make it a particularly easy-to-use and powerful development environment:

- Familiar BASIC language
- Event-driven architecture
- Object-based structure
- Extensibility

3.3.1 BASIC Syntax

The Visual Basic language is, at its core, a mature and easy-to-use version of BASIC. Even Visual Basic 1.0 contained all of the power of QuickBASIC, Microsoft's venerable BASIC for MS-DOS. The Visual Basic syntax includes all the familiar constructs that modern developers demand.

- *Looping:* for-next-step, while-wend, do-loop
- *Conditions:* if-then-else-elseif, select-case
- *Data types:* integer, single, double, long, and currency numbers, strings, variants, and user defined
- *Containers:* Variables, single and multi-dimensional arrays, structures and constants
- *Expressions and Operators:* +,-,*, -, \, Mod, Is, Like, ^, &, >, <, =>, <=, =, And, Or, Not, Xor, Or, etc.
- *Procedures:* private or public subroutines, functions, event subroutines
- *String Functions:* Len, Lcase, Ucase, Left, Right, Mid, Ltrim, Rtrim, Trim, Asc, Chr, Val, Str, Hex, Oct, Format$, Instr, String$, Space$, StrComp, Lset, Rset

3.3.2 Event-Driven Architecture

Although Visual Basic derives its language from BASIC, Visual Basic is event-driven while earlier versions of BASIC are procedural. Event-driven means that your code gets executed in response to special events. These events can come from several sources:

- *User:* Someone uses the mouse or keyboard
- *System:* A time interval elapses or a PC board initiates a hardware interrupt
- *Application:* Your code specifically calls an event subroutine
- *Other Applications:* Another Windows application requests some data or service from your application.

In contrast, a traditional procedural program calls subroutine after subroutine. In order to process input from both the keyboard and a hardware device, you must check for both devices at regular intervals -- typically in a big event loop -- and then process them appropriately. For example, the following code continually checks to see if the user pressed Q to quit or a see if a new call came in on a voice board:

```
Do
    X$=chr$(InKeys)  'Get key pressed from InKeys function
    If X$="Q" Then Exit Do
    VoiceEvent = GetLineEvent()  'Get    event    code    from
hardware
    If VoiceEvent = RINGDETECTED then
        Call ProcessCall
    End If
Loop
```

In contrast, with an event-driven language such as Visual Basic, you simply add the code in the appropriate event subroutine templates. You do not need to manage a loop or poll your sources of input for activity.

In the Code window, you select the event subroutine to which you add code. For example, the following code in bold, added to the form's KeyPress and LineEvent Event subroutine templates, accomplishes the same task as the previous example:

```
Sub Form_KeyPress (KeyAscii As Integer)
    X$=Chr$(KeyAscii)
    If X$ = "Q" Then End
End Sub

Sub Voice_LineEvent(EventCode as Integer)
    If EventCode = RINGDETECTED then
        Call ProcessCall
    end if
End Sub
```

Whenever a key is pressed in the active Form, the code will check for a Q. If a Q was pressed, the program will end. Likewise, if a RINGDETECTED Event appears in the voice board, the application processes the call automatically. Because each event has its own set of code, developing an event driven program is, by nature, very modular.

You may ask how the Voice_LineEvent Event subroutine appeared in Visual Basic. Even though Visual Basic does not include this event, this type of event becomes available if you add a third-party control for computer telephony. That is the beauty of the Visual Basic environment; if a feature is not included, you can probably purchase an inexpensive add-on to do the job quickly and easily.

3.3.3   Object-Based

Visual Basic is an object-based language. As a result, a Visual Basic application is a collection of form and control objects that you can control and query. As mentioned before, objects have their own events, properties, and methods that are only applicable to that object.

3.3.4   Extensions

You can extend the Visual Basic language in a number of ways:

- Calling C DLLs (Dynamic Link Libraries)
- Imbedding custom controls
- Communicating via Windows DDE (Dynamic Data Exchange)
- Communicating via OLE (Object Linking and Embedding)

With these extensions, Visual Basic can communicate with nearly any library or application. (Section 3.3.4 describes these terms in greater detail.) You can even combine the power of world-class applications like Microsoft Excel and Word with your own program to accomplish things that would take you years to develop on your own. For example, you can use Excel's graphing and data calculation engine behind the scenes, or programatically control Word's page layout and print capabilities. For example, using OLE and Visual Basic, you can load and fax an Excel spreadsheet named budget.xls with the following code. All you need is a printer driver that can send faxes (like the one included with Visual Fax, described in Chapter 6):

```
Dim ExcelObject As Object
Set ExcelObject = GetObject("budget.xls", "Excel.Sheet")
'Place Callers fax number in cell 4,5
ExcelObject.Cells(4,5) = CallersFaxNo
ExcelObject.PrintOut
```

3.4   Data Access

Visual Basic version 3.0 includes Microsoft's "Jet" database engine (the same engine used by Microsoft Access) and corresponding "data-aware" controls and syntax.   With these facilities, it has become possible to data-enable an application with a few clicks of the mouse.   In addition, you can fine tune an application's data access facilities at the code level with powerful new language constructs.

3.4.1   Jet Database Engine

The Jet engine allows your application to interact with a wide variety of databases in a single consistent way.   The Jet engine even allows you to have multiple simultaneous connections to different database types.   Specifically, you can access data from dBASE, FoxPro, Paradox, Btrieve, Microsoft Access, SQL Server, Oracle, and any other ODBC compatible database.

Note: Practically every database in use today has an ODBC driver available. The following is a partial list of databases available via ODBC:

Access	HP ALLBASE/SQL	RFM
AS/400	HP IMAGE/SQL	RMS
Ask	IDMS	SESAM/SQL
Btrieve	IDS	SQL/DS
CA-DATACOM	IMS	SQLBase
CA-IDMS	Informix	Sybase NetGateway
Clipper	INGRES	Sybase SQL Server
DB2	Interbase	Teradata
dBASE	MDI Gateway	Text
DBMS	Model 204	UDS/SQL
DDCS/2	NetWare/SQL	UFAS
DRDA	Oracle	VSAM
DSM	Paradox	Watcom-SQL
ENSCRIBE	Quadbase-SQL	White Cross
Excel	R:BASE	XDB
FoxPro	Rdb	

The Jet engine supports features required of modern transaction-intensive applications, such as BLOBs (binary large objects), security, and transaction services, including commit/rollback.

3.4.2   Data Controls

Visual Basic's data control is the easiest way to add database connectivity to your application. Visual Basic will automatically establish a database connection to a particular table if you simply follow the three steps below:

1. Drop a data control (located in the toolbox) onto a form
2. Specify the name of a database in the Database Property
3. Specify the name of a database table in the RecordSource Property

To display the data on the form:

1. Add to your form one or more text box or other data bound controls
2. Set their DataSource properties to the name of the data control

3. Select which DataField you want displayed in the text box from the Property window's drop down list.

When you run the application, you can click the arrows on the data control to move from record to record.

In the following figure, the data control is named Orders and is linked to the Orders table in a Microsoft Access database named ACME.MDB. The Order #, Shipped On, and Total text boxes have also been linked to the Orders table; their DataSource Properties have been set to Orders. Each control's DataField property contains the name of the corresponding field from the Orders table (i.e., OrderNum for the Order # field).

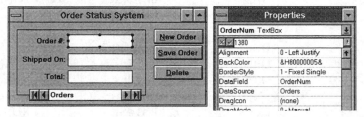

To access the data in a data control programatically, simply reference the field in the data control. The following code moves to the first record to update its total field.

```
Orders.RecordSet.MoveFirst            'Move to the first record
Total = Data1.RecordSet("Total")
Data1.RecordSet.Edit                  'Start editing the record
Data1.RecordSet("Total") = Total + 100 'Add 100 to total
Data1.RecordSet.Update                'Save the changes
```

3.4.3   Data Objects

Visual Basic Professional Edition also includes a purely programmatic way to interact with databases through special variables called database objects. Since data objects have no visual component, they can be faster and use fewer resources than Data Controls. Also, data objects allow you to create and destroy query results called Dynasets on the fly as well as create and modify the structure of a database. Of course, just like data controls, database objects allow you to perform basic data access such as stepping record-by-record through the query results, and adding, editing, or deleting records. Since all these commands are processed through the Jet engine, the same code can be used on any supported database without and modifications. For example:

```
Dim MyDB As Database
Dim MySet As Dynaset
Set MyDB = OpenDatabase("ACME.MDB")     ' Open database.
Set MySet = MyDB.CreateDynaset("Orders") 'Based on orders
table
MySet.FindFirst "OrderNum=" & ID          'Find the right order
MySet.Edit                    'Start editing the record
MySet("Total") = MyTable("Total") * 0.90    'Take 10% off
total
MySet.Update            'Save the changes
```

3.5   Other Visual Basic Facilities

Visual Basic includes numerous other graphical facilities to make application development easy.

3.5.1   Menu Editor

Visual Basic includes a menu editor so you can graphically design the pulldown menus of your application.

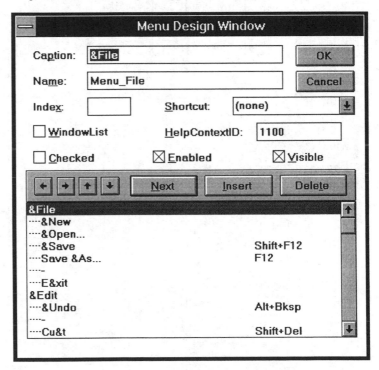

3.5.2   Database Manager

Visual Basic also includes a graphical way to design and maintain directly supported or ODBC-compatible databases.

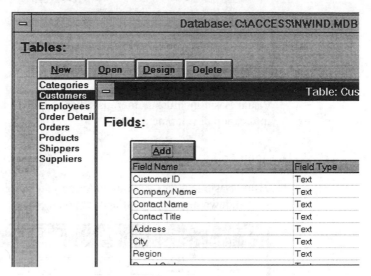

3.5.3 On-Line Help

If you ever need help on any part of Visual Basic , simply press F1
for detailed context sensitive on-line help.

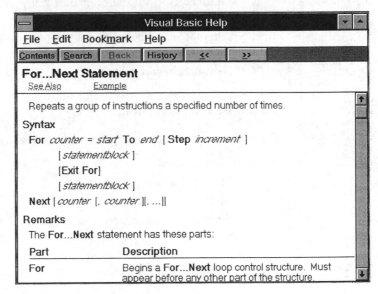

3.6 Third-Party Support

The large supporting community of Visual Basic add-on vendors
offers an incredibly wide range of products to complement the
Visual Basic foundation. If Visual Basic can not do it, you can
easily find a vendor that offers an extension that does. The
majority of third-party vendors provide add-ons in custom control
format, so you can immediately use them like any other Visual
Basic control. Practically every add-on product is reasonably
priced and most charge no runtime fees.

3.6.1 Database

Database add-ons include:

- ODBC drivers to connect to nearly any database or data format
- Alternative database engines that provide advanced features such as extremely fast execution
- Advanced report writers

3.6.2 Graphics/Image Processing

Graphics and image processing add-ons include:

- 3D animation
- Freehand drawing and annotation
- Multimedia management
- Scanning, printing, zooming, alignment, conversion, and enhancement controls for images

3.6.3 Development Productivity

Development productivity add-ons include:

- Additional reference material
- Automated testing tools
- Data repositories or data dictionaries for team development
- Documentation builders and cross reference utilities
- GUI alignment controls
- GUI installation builders
- Pretty printers and code formatters
- Profilers and tuners
- Source control and version management

3.6.4 Communications

Communications add-ons include:

- E-mail connectivity
- Faxing

- Host and mini "screen-scraping" connectivity
- Internet connectivity
- Modem communications
- Network communications
- Real time data acquisition
- Telephone connectivity

3.6.5 GUI Controls

GUI controls include:

- 3D, brushed metal, and other custom looks
- Animated buttons and cursors
- Specialized controls such as spreadsheets, tabs, outline controls, word processors, gauges, sliders, notebooks, levers, etc.

3.6.6 Other Controls

Other interesting controls include:

- Data and file viewers
- Scripting languages
- Flowchart creators
- Neural network engines and expert systems
- Numerical analysis
- Signature verification
- Spell checkers and thesauruses
- Wrappers for the Windows API, allowing you to interact with any Windows resource C programmers can

3.7 Visual Basic and Telephony APIs

With a strong, open architecture development tool such as Visual Basic, developing a telephony application is simply a matter of communicating to the appropriate telephony hardware. All telephony hardware expose certain application programming interfaces (APIs). If you can make calls to the appropriate API, you can communicate with the hardware that controls the phone lines.

Two main categories of APIs can be presented by telephony hardware and be used by Windows applications, DOS-based APIs and Windows-based APIs. Most telephony hardware that was developed before Windows became popular provides a DOS API. Some of these DOS APIs have been enhanced with corresponding Windows APIs by third parties or by the manufacturers themselves. Regardless of the API, telephony APIs under Windows need to notify the application when an event (e.g., ring detected, hang up, etc.) has occurred rather than poll the hardware constantly for new events. This is especially important in telephony applications, as telephony events can happen at any time.

Currently, telephony hardware can present one of two main Windows APIs: the Media Control Interface (MCI) or Dynamic Link Libraries (DLLs). Visual Basic allows you to call either interface, although doing so can be tricky and complex.

3.7.1   Media Control Interface (MCI)

MCI is an interface designed to handle all types of multimedia devices, from CD-ROMs, to sound cards, to digital video. Since MCI was designed to handle a variety of media devices, and voice applications have a significant media component of fax and voice, some manufacturers have decided to implement their telephony APIs using MCI. Visual Basic Professional Edition includes an MCI control that lets you send MCI command strings to any MCI device to control it. There is also a lower level MCI interface, called the *command-message interface*, which is used to send complex data and retrieve custom events from MCI devices. This interface is supported through calls to DLLs.

IBM Mwave

The IBM Mwave group has created the *Mwave* chip set with Windows drivers that are available by many different hardware manufacturers. See Section 8.2.2, "Mwave Chip Set," for more information on Mwave hardware. Mwave hardware utilizes the MCI API. Unfortunately, while you can control playing and recording files with Visual Basic's MCI control, you must use the MCI command-message interface to manage all other communication with Mwave hardware, including detecting ringing, answering calls, getting digits, detecting line drops, etc.

3.7.2   Dynamic Link Libraries (DLLs)

DLLs are sharable libraries, typically created in C, that are part of the Windows foundation. In fact, most of Windows itself consists of DLLs. As the name implies, Windows dynamically loads the code for a DLL only when the DLL is referenced; if the DLL is already loaded and being used by another application, it is shared. Every DLL can export functions, so Windows applications can call them as if they were functions in the same program.

Since DLL functions are typically defined in C format, you need to tell Visual Basic how to call them by creating a Visual Basic Declare statement for each C function. In addition, DLL functions typically have custom data types as parameters, so you need to create corresponding Visual Basic type definitions for all of the C structures expected by the DLL. Below is a Visual Basic function declaration for the Windows API function called ShowWindow, located in the Windows library USER.DLL.

```
Declare  Function  ShowWindow  Lib  "User"  (ByVal  hWnd  As
Integer, ByVal nCmdShow As Integer) As Integer
```

As shown below, calling this function can control how a Window displays at a very low level. It is not meant to illustrate how to show windows, rather it is presented to show how to make calls into C DLLs. Note that the same format applies to calling your own Visual Basic functions.

```
ReturnCode = ShowWindow(Form1.hWind, SW_HIDE)
```

Lastly, your Visual Basic application should be prepared to process special events that might be triggered by the C DLLs.

TI/F DLL

Voice Information Systems, Inc. (Santa Monica, CA) makes a Windows voice mail DLL for several boards (e.g., Dialogic, Rhetorex, and NewVoice) called the *TI/F DLL*, which stands for Telephone Interface Dynamic Link Library. This DLL includes a corresponding Visual Basic module and sample application allowing you to develop telephony applications directly from Visual Basic. The TI/F DLL makes the necessary calls to the appropriate voice board DOS driver TSR, effectively exposing a Windows DLL interface for these DOS drivers.

BICOM

BICOM provides a Windows DLL than can talk to their voice board directly. It exposes a Windows DLL interface that looks very similar to their DOS driver.

TAPI

The Microsoft/Intel Telephony API (described in greater detail in Chapter 7, "TAPI in Depth") is exposed as a Windows DLL. You can obtain TAPI via the TAPI SDK in the WinExt forum on CompuServe. TAPI will also be a native part of Windows 95 when it ships, but for now you need to install its DLL. You can also download from the WinExt forum a Visual Basic Module that exposes all of the TAPI functions and structures. To get this DLL to work with Visual Basic, you need to be notified of the TAPI events which Visual Basic cannot access directly without some help. To receive the TAPI events, you can add to your project a custom control that intercept windows messages, such as SpyWorks-VB from Desaware.

TSAPI

The TSAPI (Telephony Services API) is described by AT&T, its inventor, as a "standards-based API for call control, call/device monitoring and query, call routing, device/system maintenance capabilities, and basic directory services." It is distributed and supported by Novell and is based on Novell's very popular network operating system called NetWare. The TSAPI can be called by your Visual Basic applications through Windows via the TSAPI DLL. To allow your application to receive TSAPI events, you need to add a custom control to intercept windows messages, just as you do with TAPI.

The TSAPI communicates to Novell's Telephony Server NLM (NetWare Loadable Module), a program running on a NetWare server than handles all of the communications traffic between TSAPI and the phone system. The Telephony Services NLM communicates to the phone system via a PBX driver, a special piece of software customized for a variety of different PBXs. For more information about the Telephony Services and the TSAPI, contact Novell in Provo, Utah.

3.7.3 DOS APIs for Telephony

The majority of voice hardware vendors (e.g., Dialogic, Rhetorex, NewVoice, etc.) provide DOS TSR drivers that interact with the voice board. To call the drivers from Windows, you need to make calls to the Windows *DOS Protected Mode Interface* (DPMI). This is required because DOS TSRs and programs run in real mode, while Windows runs in protected mode. Calling through this API is quite complex; see the Microsoft Windows SDK for more information on using DPMI.

CHAPTER FOUR

Telephony Concepts

4.1 How a Telephone Circuit Works

Your telephone is a simple device that allows you to communicate over the world's most advanced network by pressing only a few digits. With this device, you can talk with anyone else in the world who also has a telephone. How does your phone connect to this amazing network? This section discusses how.

4.1.1 Your Telephone and the Local Loop

An ordinary, analog telephone (also called *POTS*, or Plain Old Telephone Service) is connected by wire directly to the telephone network at the *Central Office* (or CO for short). The circuit between your phone and the CO is called the *local loop*. If you have ever heard the terms *tip* and *ring*, they simply refer to the two wires in the phone cord, connecting to the local loop.

To initiate an outgoing call or answer an incoming call, you pick up the handset. The switch in the telephone cradle closes and the telephone goes *off-hook*. This causes current to flow through the telephone and requests service from the Central Office. When you place the receiver down to end a call, the telephone is said to be *on-hook*.

4.1.2 T1 Digital Lines

In larger phone installations, your phone lines can be transmitted from the CO to your system via a *T1 trunk* that carries 24 telephone lines, or *channels*, in one phone cable. These T1 lines are digital and need a special interface to the voice board, but most multi-line voice boards have this option. You must also make sure that the software you use supports the protocols involved with T1 communications.

4.1.3 PBXs

Sometimes a telephone line is not directly connected to the CO, but is instead connected through a *Private Branch Exchange* (PBX) or *key system*. You may find one of these telephone environments in your business organization -- a PBX acts somewhat like a company's own private Central Office. The PBX is connected, either by several analog lines or a digital trunk, to the nearest Central Office.

Calls can be forwarded and transferred within the corporation with special signaling. Each brand of PBX uses its own form of signaling, and some voice boards enable you to detect and control this signaling.

Many PBXs are digital, which means that calls within the company are transmitted digitally, rather than with analog signals. The telephones used with these digital PBXs are proprietary and are not the same telephones that you would find connected to a POTS line. This also means that an ordinary voice board cannot connect directly to such a PBX. You have three options:

- Bypass the PBX and connect to a POTS line directly from the Central Office.
- Purchase and install a *two wire analog station card* into the PBX.
- Buy a voice board customized to connect to the digital lines of your PBX (not available for all PBXs).

4.2 Establishing a Connection

We have described how telephone calls are switched from one telephone office to another, but how does one actually establish the connection?

4.2.1   Receiving a Call

When someone calls you, you know to pick up the handset because the telephone rings. (This ringing is caused by a momentary spike in the voltage across the phone.) When the handset is lifted, current flows through the telephone and the connection is established.

If your organization has several inbound telephone lines, you can have the telephone company set up a service called *hunting*. This causes a phone call dialed into Line #1 to bounce to the next line if the first line is busy. With hunting, you can publish one phone number but handle multiple phone calls at one time. This feature should be transparent to your computer telephony system, which treats each phone line separately, and should be transparent to your callers, who always call the same number.

4.2.2   Placing a Call

To initiate a telephone call, you must lift the handset. This causes current to flow from the CO through the phone and indicates that you desire telephone service. The Central Office sends you a signal (the *dial tone*) to indicate that you can proceed with the call.

Two methods of dialing exist: touch-tone and dial-pulse. Touch-tone dialing is very popular because it is quicker and more accurate, and it is useful for interacting with voice response systems. While touch-tone dialing is not entirely widespread yet (older-style rotary telephones use dial-pulse dialing), it is estimated that 70% of homes in the United States have touch-tone service.

Call Progress Analysis

Once a telephone number is dialed, the Central Office attempts to complete the call. At this point, one of several things may happen. You might hear:

- Nothing. If there is no ring-back or other signal detected, the number was probably incorrectly dialed. This will probably be followed by a reorder tone.
- A ringing with no answer, which is an indication that the telephone you are trying to reach is ringing on the other end, but no one is answering the telephone.

- A busy signal, which indicates that the other telephone is in use.

- A fast busy signal, which means that the interoffice trunks are busy, and you should try back again.

- An *operator intercept*, which is a voice message preceded by three rising tones, such as, "boop-Boop-BOOP...The number you have reached, 555-1212, is not in service at this time..."

- A *reorder tone*, such as a dial tone, prompting you to try again and "reorder" the phone number or service. This generally indicates the phone system timed out because a valid telephone number (or other request) was not entered quickly enough.

- A fax receive signal, indicating that a fax machine answered.

- Ringing then a connect followed by a voice. This means the call was answered.

A connect can be detected in many ways, and some voice boards can even detect the sound of someone answering the phone to determine that a connection was established. The telephone company uses a certain set of *call progress signals* to indicate call status, and most voice boards understand these by default. If you plan to do automated outbound dialing, you should select a voice board that performs reliable call progress analysis and can be trained to understand your particular phone system's signaling.

For some systems, the call progress signals are different, and the voice board must be adapted to the new signaling.

Flash Hook

Sometimes one party, while connected to another party, wants to request a service from the Central Office or PBX. Requesting such a service is usually initiated with a signal called a *flash hook*. Sometimes the flash is followed by touch-tones to indicate which service is desired. Two common examples are:

- *Call waiting:* the recipient of a second call, while on the phone with the first caller, can issue a flash and be connected to the second call.
- *Call transfer:* a receptionist can transfer a call to another extension on the PBX with a flash followed by the extension to which the call should be transferred.

A flash puts the phone on-hook for just long enough to signal the request, but not long enough to disconnect the call.

4.2.3   Disconnect

At the end of a call, one of the parties puts the receiver on-hook. Usually, a *disconnect signal*, in the form of a drop in loop current or a disconnect tone, is then transmitted to the other party.

With some systems, the disconnect signal is preceded by a delay of many seconds, and in some situations it may not be transmitted at all. This is a special consideration when integrating with a PBX, since many PBXs do not transmit the disconnect signal. If you already have such a PBX, you may be able to connect your IVR system to the PBX's serial port to gather this information. Otherwise, you may not know when the caller has hung up and will have to rely on "timing out" when there is no response from the caller in order to ensure each call is terminated.

4.3   Advanced Telephony

Some applications require advanced features, such as switching, voice recognition, Text-To-Speech, and call identification.

4.3.1   Switching and Conferencing

For transferring and conferencing calls without a PBX or sharing a single fax board between multiple voice lines, your application will need to switch calls between phone lines. This requires a switching board and special software.

With switching support, your application can:

- Connect calls to other telephony devices (*e.g.*, to connect an incoming fax to a faxmodem)
- Share devices among several voice lines

- Ring (or blind-ring) any telephone or device connected to the board and wait for it to pick up (*e.g.*, transfer a call to a live operator)
- Connect two (inbound or outbound) telephone connections together (*e.g.*, *international callback*)

You can connect any switch board port to devices like:

- Central Office (outside telephone lines)
- Voice board ports (lines)
- Telephones (connected through ring generators)
- Fax machines and modems
- Other telephony-related hardware (*e.g.*, voice recognition cards)
- Music-on-hold

Two primary types of boards exists: analog (e.g., Dialogic AMX) and digital (e.g., Dialogic DMX /MSI or Dianatel).

The AMX (Analog MatriX) board is an 8x8 matrix of ports with X and Y axes that you can switch together. You can connect voice boards via the AEB (Analog Extension Bus) to channels along one axis, attach other devices to channels along the other axis, and then selectively switch them together. The diagram below illustrates several possible connections with an AMX board, simple fax modem, CD player, and voice board.

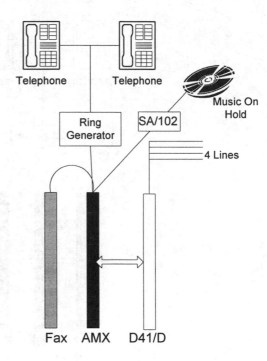

One disadvantage of the AMX is its substantial volume loss when switching together any two external voice ports. Connected callers will notice quieter voices, and touch-tone digits may potentially not be recognized by the voice board in such a configuration. This signal loss does not effect local phones connected to the switch or other audio devices.

Digital switching boards work in a similar way, except they provide many more connections (up to 96 with the DMX/MSI and up to 196 with Dianatel), are more flexible (ports can be connected either one- or two-way to any other port), and can do it all without signal loss. When evaluating digital or analog switches, make sure you understand your current and future application needs, as some features like conferencing, ringing attached phones, and detecting signaling such as line drops may require additional hardware. The following is a diagram illustrating how devices can connect to a Dialogic DMX/MSI combination. Note that since these switches communicate over the digital PEB, you cannot connect AEB voice boards (i.e., D/21x or D/41x), you must instead use voice boards that can connect to the PEB such as the D/121B and D/240SC.

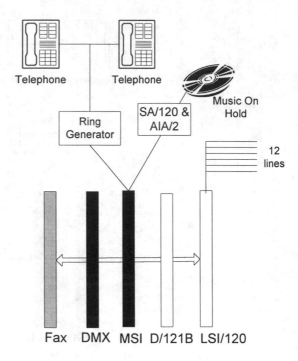

Refer to Section 8.5, "Switching Solutions," for more information about switching hardware.

4.3.2 Text-To-Speech

Most computer telephony applications have some sort of rudimentary *Text-To-Speech* capabilities for speaking variable data, such as numbers, dates, and currency. However, pure Text-To-Speech, or TTS, that can read e-mail messages, addresses, or names is a technology that is only recently gaining wider acceptance.

Depending on the type of TTS technology used, TTS can sound artificial. In general, if the information can be broken down into a finite number of words or phrases, pre-recorded voice files sound better. However, TTS is essential for applications that require reading raw ASCII text such as addresses, articles, messages, or names. Additionally, TTS can reduce disk storage space by two orders of magnitude when the number of required voice files would otherwise be large, and is worth considering where quality is not a primary concern.

Text-To-Speech Integration Methods

You can obtain Text-To-Speech using three methods: software, firmware, and hardware.

Software-only products load the algorithms into your PC's CPU, which performs all of the processing. These products are generally less expensive because they do not require any hardware. On the other hand, they are slower and reduce the number of lines your PC can handle, because they drain its resources.

Firmware products can be loaded onto your voice board's processor, which reduces the load on your PCs CPU and does not require additional hardware either. If you need this feature, check with your hardware manufacturer to make sure they offer an acceptable firmware Text-To-Speech option.

TTS hardware requires you to add an additional board into your PC and share it with your voice lines via a switching board or voice bus.

See Section 8.6.1 for a list of TTS solutions.

Techniques

Theoretically, a Text-To-Speech system takes a section of ASCII text and converts it into human speech. In reality, TTS does not sound like perfect human speech because of the variety of exceptions and situations the system will encounter.

A good Text-To-Speech system first determines the context of any ambiguous numbers and abbreviations. Next, it looks up each word in the dictionary to see if any special pronunciation is documented. Third, it converts the words into phonemes -- small units of sound that can be combined to form full words. A more sophisticated TTS system may modify the pronunciation and inflection of certain syllables depending upon words next to the word or based on the context. Finally, it generates the voice.

Features

First and foremost, when you select a Text-To-Speech solution, you should consider the quality of the pronunciation. Can you clearly understand the terms, names and addresses you commonly use?. This depends on both the algorithms for analyzing the text and the sound of the voice. Listen to the product speaking phrases that are common to your applications. For example, applications that require speaking proper names and addresses are particularly challenging to Text-To-Speech systems. Other features to keep in mind include:

- Foreign languages and voices available
- Ability to change volume, speed, pitch, inflections within sentences, etc.
- Integration method (software, firmware, or hardware)
- Number of ports available

4.3.3   Voice Recognition

Voice recognition, also called VR or speech-to-text, enables your system to understand words spoken by a caller. Voice recognition is distinguished from speech recognition, where the telephony system employs pattern matching to identify the caller. Voice recognition identifies the words.

Computer telephony systems have begun capitalizing on VR to reach phones that do not have touch-tone service. Voice recognition enables callers to speak numbers instead of pressing them on their keypads. (See also Section 4.3.5, "Pulse Dialing" below for another option.)

Voice recognition technologies for telephony are currently available in speaker independent, discrete and continuous recognition versions. Speaker independent means that potentially anyone can be understood by the VR system without training or a particular users. Discrete recognition forces the speaker to pause briefly between each word spoken, while continuous allows users to speak words without pauses. Compared to speaker dependent VR performed over a desktop microphone that can recognized thousands of words for dictation applications, telephony based VR can typically only recognize between 10-60 words because of three main quality factors:

- the 8000hz bandwith of voice data over the phone does not provide enough clues for highly accurate pattern matching
- the quality of caller's handset microphones and related telephony hardware varies widely

. additionional background noise can be present further
reducing quality

VR technologies come with one or more recognition vocabularies.
Vocabularies typically include international languages, simple
touch-tone replacement (e.g., "one", "two"), spelling, voice mail
commands (e.g. "Play", "Rewind", "Delete"), and other
application specific templates. Usually custom vocabularies can
be developed for specific applications.

Your computer telephony system can achieve voice recognition
through software or hardware, as described in the previous section
on TTS. For a more detailed account of available solutions, refer
to Section 8.6.2, "Voice Recognition Solutions.".

4.3.4 Call Identification

The phone company can configure telephone lines so that when an
incoming call rings, the Central Office or PBX sends extra
information to identify the call. In every case, you must have
hardware and software that supports the service. Four types of call
identification services are available:

. *Automatic Number Identification (ANI):* On 800 lines or
special PBXs, ANI tells you the phone number from
which the phone call originated. This service is provided
on T1 (digital) lines via *MF* (multifrequency) tones or
regular touch-tones. This service is useful for
determining who is calling.

. *Caller-ID:* A service similar to ANI, but offered by the
local phone company. Availability of this service
depends on each state and the local telephone company.
The FCC has currently mandated that all phone
companies must provide Caller-ID by May 1995, but the
actual availability of service in your area may be later.

. *Dialed Number Identification Service (DNIS):* On T1
lines, DNIS provides you with the number that the caller
dialed. This information is sent by the phone company as
touch-tones. DNIS is useful to determine which number
the caller dialed, so you can start the appropriate
application or route the call appropriately.

. *Direct Inward Dial (DID):* DID is the analog equivalent of DNIS.

To receive ANI, DNIS, or DID information, you must perform a *wink* to rapidly go off-hook, receive the information, and then go on-hook again. The wink must be long enough to receive the information, but no too long so the Central Office thinks you want to answer the call yet.

4.3.5   Detecting Special Signaling

Most telephony systems can detect touch-tone signals by default. However, some PBXs and key systems use special tones (as opposed to touch-tones) to communicate special information. For example, your PBX may emit a custom tone when you have another incoming call. Certain voice boards and software provide custom tone support that enables you to define tones that you can play or detect at any time.

Your application might have to detect signaling such as:

- Pulse dialing
- Fax send or receive tones
- Special telephone system signals
- Custom-defined call-progress signals

Pulse Dialing

As many as 30% of households in the Unites States, and even more internationally, have only rotary-pulse dialing service. If you need to reach these households, you must offer another data entry method besides touch-tones. As mentioned earlier, voice recognition is one option. An easier solution is to simply purchase a pulse-to-tone converter that translates any incoming pulses into touch-tones and transmits them to the voice board.

Fax Send and Receive Tones

Another feature to consider is voice/fax discrimination. A fax machine sends a special signal when it is ready to either send or receive a fax. If you configure your system properly, it can recognize a fax machine and then either hang up, receive the fax, or transfer the call to a fax machine or another application that can receive the fax. Most voice/fax/modems have this feature built-in, while most professional voice boards provide custom tone detection capabilities that you can train to detect these signals.

Custom Tone Detection

Some professional voice boards and software can detect special tones using custom tone support. This support is meant to supplement the telephony application's ordinary touch-tone detection. Practically any special tone or tone pattern can be recognized at any time during a call. Examples include:

- Fax send and receive tones (see above)
- Modem send and receive tones
- Special line dropped signaling
- Call waiting tone
- Reorder tone

Custom tone detection can also be used to fine-tune the detection of "call progress" signals (*e.g.*, busy signal, dial tone, ringing, *etc.*) that occur right after you place a call. This is covered in the next section.

You can use custom tone support to communicate with PBXs that use *in band signaling*, a method of communication where tone patterns are played over the telephone connection. Some PBXs and key systems use *out of band* signaling instead, a method where all communications occur separate from the telephone line. Sometimes your application can access out of band signaling via a serial connection between the PC and the PBX's switch (or via a special data channel in digital systems). Signaling over a serial connection can be controlled with Visual Basic's serial communication control, MSCOMM.VBX.

Custom-Defined Call Progress Signaling

Not all telephone systems use standard call progress signaling. If your voice board is having trouble placing outbound calls and if it supports custom tone detection, you can fine-tune your call process algorithm.

Professional voice boards often include programs that will automatically learn your phone system's call progress signaling.

You can usually use the characterization program to also manually learn other custom tones, as described in the previous section.

4.4 General System Issues

A basic computer telephony system consists of a PC with the appropriate software and voice response hardware connected to your phone system.

4.4.1 System Requirements

The following are the minimum system requirements for running computer telephony applications with Visual Basic.

Hardware

- 386, 486, or Pentium PC with:
    - At least 4 Mb of RAM for 2 lines (more memory is required for 4 line and higher applications)
    - Adequate hard disk space for voice file storage (See "Voice File Storage Requirements" in Section 5.2.3 for a table of storage requirements.)
- To play and record voice files, one of the following:
    - A professional voice board
    - A voice/fax/modem
    - Some other form of telephony-enabled voice hardware

    See Chapter 8, "Telephony Hardware" for a list of manufacturers

- (Optional) A fax board for faxing
- One or more telephone lines connected to your voice board port(s)

Software

- Microsoft Windows™ or Windows For Workgroups™

- Microsoft Visual Basic runtime DLL

- Your telephony application, including all of its voice files and databases

- A driver for your voice board.

Additional Requirements for Developing Applications

The following are additional requirements for developing your own computer telephony applications in Visual Basic:

- Microsoft Visual Basic development environment

- A toolkit, dynamic link library, or API for interfacing to the voice board driver from Visual Basic

- *For testing:* Some environments provide a software phone to simulate phone conversations using any SoundBlaster®-compatible device. However, for systems that do not provide a software phone, you will need a real phone and one of the following two configurations:

  - Two phone lines, one for the phone and another for the voice board, that can call each other through the phone company
    - *or* -
  - A line simulator connecting the phone and the voice board together to simulate a telephone connection.

Note: Line simulators make it easy to test and demonstrate your telephony applications. By providing a direct connection to the voice board, a line simulator allows you to simulate the basic functions of a Central Office during application development. As a result, you can achieve consistent and reliable testing while reducing the need for additional outside telephone lines. Refer to "Line Simulators" in Chapter 8 for a list of line simulator vendors.

4.4.2 Selecting Your Voice Response Hardware

When selecting a voice board for your telephony application, you should consider a variety of factors:

- Number of ports, or lines, supported
- Chip architecture
- Voice bus interface
- Telephone line interface
- Support for TAPI
- Telephony features supported

Number of Ports Supported

Voice processing boards vary in the number of *ports* or *channels.* This number indicates simultaneous phone calls supported per board. Two primary categories of voice boards exist:

- multimedia voice/fax/modems
- professional telephony-only voice boards

Multimedia voice/fax/modems generally support only one telephone call at a time (though some are appearing in two-line versions). In most cases, only one of these boards can be installed in a single PC. These boards are primarily positioned to reach the SOHO (small office / home office) market. These voice boards usually also provide fax, modem, and other capabilities at a fairly low cost. Consequently, these boards offer great value for single-user telephony applications.

Professional voice boards, on the other hand, are solely dedicated to voice processing and telephony functions. These boards are high-quality, robust boards that can handle anywhere from two to thirty phone calls at one time, depending on the model. Generally, several multi-line boards can be installed in a single PC to further increase the capacity of the system. While these boards often cost more, they are the appropriate choice for multi-line or mission-critical applications.

DSP Chip Architecture

Until a few years ago, voice boards were limited in lifespan because customers would have to buy new hardware any time a new feature would become available. Over the last few years vendors have begun providing voice boards based on digital signal processing (DSP) technology. Vendors can immediately add new features to DSP-based boards through software upgrades to the board's firmware. These generally free or inexpensive firmware upgrades increase board longevity.

Voice Bus Interface

A voice bus enables you to extend your existing telephony system via the ability to pass audio/voice information between additional voice processing boards and related hardware (e.g., fax boards, voice recognition hardware, Text-To-Speech components). Voice processing hardware vendors have developed various voice bus standards:

- *Analog Expansion Bus (AEB):* Dialogic Corporation's proprietary analog bus.
- *PCM Expansion Bus (PEB):* Dialogic Corporation's proprietary digital bus.
- *Multi-Vendor Integration Protocol (MVIP):* A digital bus created by a consortium of vendors to allow cards from multiple vendors to communicate.
- *Signal Computing System Architecture (SCSA) bus:* The digital bus that is part of Dialogic Corporation's open standard for interconnectivity between telephony components. The SCSA bus differs from the other digital bus standards because it can also support communications between several PCs over a LAN. Like MVIP, the SCSA bus is an open standard; the specifications are available to anyone developing components for the SCSA architecture. Since its introduction in early 1993, the SCSA standard has been endorsed by over a hundred software and hardware manufacturers.

Telephone Line Interface

Voice boards support many types of network services, the most common in the U.S. being POTS and T1 (others include E1, ISDN, etc.). Higher density boards that connect via the PEB, such as the Dialogic D/121B and D/240SC, separate the telephone interface from the voice processing component so that you only need to replace the line interface card to change from an analog to digital interface at a later date.

Tone dialing is always directly supported by voice boards. In certain areas where pulse or rotary dialing is used, additional hardware may be needed. See Sections 4.3.3 and 4.3.5 for further details.

Support for TAPI

The new Microsoft/Intel Telephony API standard enables developers to build applications that will automatically work with any TAPI-compliant hardware. If hardware independence is important to you, consider creating a TAPI-compliant application and make sure that the board you select offers a TAPI service provider that interfaces to your software. Refer to Chapter 7 for a complete description of the Telephony API.

Advanced Telephony Features

Voice boards are designed to detect telephony signals, understand touch-tone input, and play and record voice files. As covered in Section 4.3 above, some voice boards support enhanced telephony features including call identification (e.g., Caller-ID, DID, DNIS, ANI), call progress detection, and global tone detection. If you plan to implement any of these features, make sure that your hardware supports them.

Fax Integration

Some voice boards also include on-board fax modems or support for fax daughter boards. This leads to easy integration of fax-on-demand applications. Otherwise, you must purchase fax boards separately.

4.4.3   Selecting Your Fax Hardware

A variety of fax modems abound in the marketplace today and the available features vary greatly among vendors. The wide range of fax boards encompasses internal and external boards as well as hybrid voice/fax boards and fax daughter boards.

When selecting fax hardware for your telephony application, you should consider a variety of features:

- Baud rate
- On-board processing
- Number of ports supported

- Voice board compatibility
- Transmission protocols
- Fax API type
- File formats supported

Baud Rate

The speed, or baud rate, at which fax transmissions take place will have a direct effect on operating costs (i.e., phone call charges, number of phone lines required) as well as overall performance of the application. Baud rates, measured in bits per second, commonly range from 9600 bps to 14,400 bps.

On-Board Processing

Central to system performance is whether a fax board has an on-board processor to service the fax or if the processing is to take place on the host CPU. If the card has a processor directly on the board itself to handle the faxing, it allows the PC to handle other non-faxing processing. This results in maximum system performance.

Number of Ports Supported

Some fax cards are single-line, while others can support up to 32 lines in a single card. As with professional voice boards, multiple fax boards can usually be combined in a single PC. For high-volume faxing applications, you should carefully estimate the number of phone lines needed.

Voice Board Compatibility

For fax-on-demand applications, you should consider how you will integrate the voice and fax. Voice and fax integration can be achieved in several ways. At the hardware level, some manufacturers provide direct support for voice and fax on a single board.

Alternatively, you can combine separate voice and fax boards in a single PC. If you will share fax boards between voice processing applications, you should consider switching support or fax and voice boards that can communicate over the voice bus.

Transmission Protocols

Fortunately, the majority of modern-day fax machines conform to the same standard (T.30 fax transmission protocol or Group III standard). Most manufacturers will have a compatibility list for verification.

Fax Modem Types

For the purposes of this book, there are four main types of fax modems: Class 1, Class 2, CAS, and high-density.

Class 1 and Class 2

Class 1 and Class 2 modems are probably the most popular type of fax modems. They communicate via the serial port for all fax activity and hence are limited by the bandwidth of serial ports and the PC processor's ability to transfer data over the serial port. Class 1 was created by the ITU-T (International Telecommunication Union - Telecommunication sector) in 1990 and was the first to use Hayes AT commands to control faxing activity.

Class 2 was the next *proposed* standard that never passed due to political issues but unfortunately was widely adopted anyway by some major manufacturers. The Class 2.0 specification was finally approved by the ITU-T. Unfortunately, this standard was approved so late that it does not include many features of the newly approved T.30 protocol, such as better handshaking.

If you plan to plan to use Class 1, Class 2, or Class 2.0 fax modems, be careful of the numerous compatibility and other issues associated with them.

CAS

The Communicating Applications Specification (CAS) API was jointly developed by Intel and DCA.

Fax modems using CAS, such as the SatisFAXtion modem, provide a queue-based interface. This enables an application to submit fax files to a queue for processing.

Typically, CAS-based fax/modems have a microprocessor on the fax/modem itself, relieving the CPU of fax conversation and image management. CAS allows scheduled faxes, binary file transfers, and the ability to overlay text and graphics on a single page (i.e., for cover page or other merging applications).

High Density

High density fax boards include those from GammaLink, Brooktrout, Dialogic, etc.

Because these fax boards typically provide four or more lines per board, you can easily support sixteen or more fax lines per PC. Superior outbound call progress monitoring and error handling allows your application to detect various line conditions such as operator intercepts, no ringing, etc. For example, GammaLink provides nearly 600 extended status codes to help you determine what went wrong when a fax failed to send.

Specialized interfaces allow you to hook up directly to T1 or E1 lines or voice buses such as the AEB, PEB, or MVIP. By connecting a fax board to a voice bus, you can connect all your phone lines to the voice board channels and then share your fax boards across the telephony bus when a fax service is requested.

Fax File Formats

Depending on the type of support you use, you can send and receive faxes in a variety of formats. Common sending formats include TIFF, PCX, DCX, and ASCII. Most fax devices receive fax files in DCX format. Regardless of what fax formats a particular device supports, you can use a variety of conversion libraries to convert your documents to or from any valid fax file format. In addition, most types of fax support include a printer driver so you can print documents from any Windows application into a fax-able file.

4.4.4   Telephony System Configurations

Computer telephony systems range from single-user desktop dialers to call center systems that can handle hundreds of calls at once. Each system is configured differently, but this section describes two common examples.

Stand-Alone System

A stand-alone system interfaces directly to one or more POTS lines, as shown below.

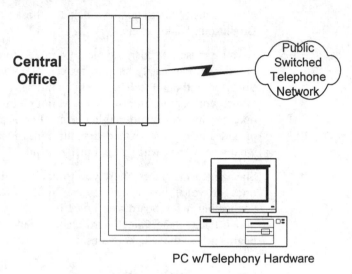

PC w/Telephony Hardware

This type of system can both answer inbound and place outbound calls.

To build this type of system, you would insert one or more voice boards into a PC and connect them directly to the telephone line. (The connectors on the back of the board are identical to the jacks found on standard telephones. Four-line boards sometimes only have two jacks -- each jack accepting a two-line cord.)

You can run different applications on each line. If, on the other hand, you are running the same application on all lines, you can choose to only publish the phone number for Line #1 and have calls hunt to the other lines when Line #1 is busy (refer to Section 4.2.1, "Receiving a Call," for more information about hunting).

If your system handles 24 (or more) phone calls at once, you can install a T1 digital trunk to bring all 24 lines onto the premises in a single cable. Some voice boards, such as the Dialogic D240SC-T1, can interface directly to T1 lines or through a network interface card such as the Dialogic DTI-211. . Otherwise, you can install a *channel bank* to convert the digital trunk into 24 analog lines.

In international markets it is common for 30 calls to be delivered at once over E1 digital lines. Just like with T1 lines, there are similar ways to connect voice boards to E1 lines.

Integrating your System with a PBX

For typical voice mail and auto attendant applications, a computer telephony system utilizes the call control features of a PBX to transfer, conference, and hold calls. Typically, the auto attendant would be connected to one or more extensions on the PBX, as shown below.

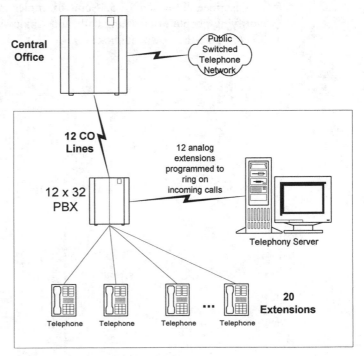

The PBX should be programmed so that all inbound calls are first routed to an auto attendant extension. (Alternatively, it could attempt a receptionist's extension first, so that the caller only encounters the auto attendant if the receptionist is not available.)

The auto attendant then prompts the caller for the extension and then tries the extension in the same way the receptionist would. Typically, this involves pressing the transfer button (i.e., issuing a flash-hook) and dialing the extension. If the other party answers, the auto attendant hangs up, transferring the call. Otherwise, the auto attendant issues another flash-hook to return to the calling party and prompts the caller for a message. (Refer to Section 5.2.2 for more information about transferring calls.)

Because PBXs usually use proprietary signaling, this type of system requires voice boards that have excellent, customizable call progress detection (and, in some cases, the ability to play custom tones). Refer to Section 4.3.5 for more information about detecting special signaling.

In companies or call centers that have computers connected on a LAN, the auto attendant can offer very slick client/server features, as described in Section 5.1.5. Some examples include screen pops notifying people of incoming calls and windows that enable users to configure their own mailboxes.

Developing Telephony Applications

5.1   Examples of Computer Telephony

Computer telephony applications can be classified into five main categories:

- *Call processing:* voice mail, auto attendant, ACD
- *Voice response:* audiotex, IVR, data entry
- *Outbound dialing:* predictive dialing, automated outbound dialing
- *Fax processing:* fax broadcasting, fax-on-demand, fax server, fax mail
- *Client/Server:* network-based call control, message retrieval, etc.

While the following sections discuss each category separately, most applications actually combine the above technologies.

5.1.1   Call Processing

Call processing systems answer inbound calls and route them appropriately. The call can be routed to voice mail, a phone extension, or the next available sales representative.

Voice Mail

Voice mail is so ubiquitous today, most people use it on a daily basis. Simple versions require only the ability to record voice files, save them in a database, and then play them back when retrieved. Most versions save the time and date with the message and enable the user to selectively listen to, delete, and save messages. More advanced versions allow listeners to rewind and fast-forward through messages, forward messages to other users, change the outgoing message, and more.

Auto Attendant

An *auto attendant* automatically transfers a caller to the appropriate phone extension based on touch-tone entry. Most auto attendants let a caller either type in the extension or use touch-tones to spell the desired party's name. A peripheral to a company's phone system (see "Integrating your System with a PBX" in Section 4.4.4 for a sample configuration), an auto attendant almost always mimics a receptionist to request the appropriate extension (for example, pressing the transfer button followed by the digits of the extension).

ACD

An *ACD* (Automated Call Distribution) application usually transfers incoming calls to the first available operator or salesperson. These types of applications can use logic as simple as, "Try extension #1. If busy or no answer, try extension #2. If busy or no answer, try extension #3, etc." Alternatively, these applications can be quite complex and route calls all over the country to find available operators.

An ACD can first determine the caller's identity or reason for calling and route the call appropriately. Sophisticated systems will cause the corresponding screen of information to appear on the operator's workstation. For example, a service bureau will offer 800 service to various catalog retailers, who each publish a different 800 number. Based on the number dialed, the operator answers with the appropriate retailer's name and refers to the database of product information to answer questions.

5.1.2   Voice Response

Voice response systems speak prerecorded messages and information to callers. Voice response systems can be categorized into three types: audiotex, IVR, and data entry.

Audiotex

Audiotex (also referred to as "audiotext") allows callers to access pre-recorded messages through a fixed set of menus. This information is not updated in real-time. For example, a caller can enter the name of a commuter rail station and hear the departure times from that location.

Other examples include:

- Movie show times
- Special event information
- Store hours
- Club bulletin board
- Talking classifieds
- Weather hotline

IVR

IVR, or Interactive Voice Response, can speak a wide variety of information directly out of a database over the phone. This information is generally real-time data, or is at least updated very frequently.

Because the information is stored in a database, callers can tailor the spoken information to their needs. For example, callers can access their bank's touch-tone banking system, enter their account numbers, and hear their bank balances. An IVR system can speak a wide variety of caller-specific information, including:

- Order status
- Inventory levels
- Credit card balance
- Benefits status
- Mortgage payment amounts

Data Entry

Callers can also enter data into a voice response system's database. The system can confirm the caller's input after each entry. Examples include:

- Order entry
- Phone survey
- Field repair crew materials reporting
- Remote time and attendant tracking
- Work orders entry
- Benefits enrollment

5.1.3   Outbound Dialing

Outbound dialing systems automatically place phone calls and either route the call to a live operator or leave a message.

Predictive Dialing

Predictive dialing maximizes the efficiency of outbound telemarketing/telesales representatives. The system dials the phone number of each prospect and determines whether a human answers. If a person is detected, the system immediately switches the call to an available telemarketing/telesales rep.

Automated Outbound Dialing

An automated outbound dialing application involves no human telemarketing/telesales representatives. Instead, the computer plays a specific message once an answering machine or person picks up the phone. Examples include:

- Reminding patients of appointments
- Updating voters about campaign issues
- Reminding constituents to vote
- Announcing last-minute meetings
- Paging sales reps on the road

5.1.4 Fax Processing

A fax processing system can send information instantly in written form, extending the possibilities of any computer telephony application. This section describes four types of fax processing applications: fax broadcasting, fax-on-demand and fax store-and-forward, and fax server.

Fax Broadcasting

A fax broadcasting system automatically sends a document to a list of fax numbers in a database. This type of system is valuable for applications such as:

- Newsletters
- Late-breaking event information
- Product upgrade notifications

Usually, a system will wait until after-hours to fax the documents in order to save on toll charges. Some systems will customize the cover page, or even customize the contents of the fax, based on the recipient.

Fax-on-Demand

A fax-on-demand (FOD) system lets callers request information via fax documents. A caller can use touch-tones to select certain documents to be faxed

A *two-call* FOD system asks for the person's fax number and calls back. A *one-call* FOD system sends the document on the same call. The one-call system requires that the requester call from a fax machine, and it places the burden of telephone charges on the caller.

Common applications for fax-on-demand include:

- Product literature
- Technical support information
- Real estate listings

Most systems simply fax existing documents; however, some systems can actually compile customized reports for the caller.

Fax Mail

Fax mail, or fax store-and-forward, enables callers to fax documents to private fax mailboxes. Owners of the fax mailboxes can then retrieve these documents from any other fax machine. This system is particularly beneficial to sales reps on the road who could not otherwise receive their faxes. Fax mail also offers privacy that an ordinary fax machine does not provide.

Fax Server

A fax server sends faxes at the request of any user on a LAN. Generally the fax hardware is centrally located at one node. This arrangement gives all users in a company access to fax capabilities without requiring a fax modem in every PC. As an added benefit, the server can be programmed to send the faxes late at night when phone charges are lower.

5.1.5   Client/Server Applications

Some of the truly exciting telephony applications to be developed involve many client workstations communicating to a telephony server. Often called client/server telephony, this arrangement usually employs Windows workstations because the vast majority of end users are running Windows as their operating platform.

Client/server telephony can provide everyone on the network a graphical user interface to their voice mail and auto attendant system. Essentially, one server on the network contains the voice response hardware, and other nodes communicate with the server to handle inbound phone calls and make outbound calls.

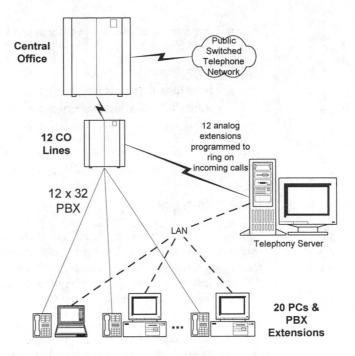

Two exciting applications for client/server telephony are call screening and universal messaging.

Call Screening

A call screening application notifies users when incoming calls are being routed to the users' phones. To determine the caller's identity, the call screening application can either use Caller-ID, ask the caller to speak his or her name, or ask the caller to enter an ID number. The system alerts the user of an inbound call with a screen pop,. In the first scenario, the screen pop notifies the user of the caller's phone number and name if a match is found in the Caller-ID database. In the second scenario, the auto attendant sends a .WAV file of the caller's name over the LAN and "whispers" the name over a sound device (e.g., SoundBlaster card) in the user's PC.

The call screening application can offer several options to the user, such as:

- Take the call now
- Send the call to voice mail
- Route the call to another extension
- Put the caller temporarily on hold

The network's communications and dialog boxes make Visual Basic particularly well suited for this type of application.

Unified Messaging

Unified messaging presents e-mail, voice mail, and fax mail to the user with one consistent GUI that closely resembles existing e-mail messaging screens. Just as users today click on e-mail messages to display them, users click on voice mail messages to hear them and click on faxes to view them.

This type of graphical interface allows users to immediately identify the number of voice mail messages waiting, the duration of each message, the time each message arrived, the urgency of each message, and the sender's identity. This information makes it much easier to manage and prioritize voice mail messages.

Visual Basic simplifies the development of these types of applications because Visual Basic can handle connections to Microsoft Mail via MAPI, and custom controls are available that can link to alternative e-mail packages.

5.2   Elements of a Telephony Application

Following are some basic guidelines for structuring a computer telephony application. Most voice boards and their APIs support the functionality described in this section. When you select your system components, make sure they support the features you need.

5.2.1   Answering Inbound Calls

Generally, when a voice board detects an inbound call, an event is generated which indicates that the phone is ringing. Your application should automatically pick up the line. You should specify how many rings the system should wait before answering the phone.

As described in Section 4.3.4, your application can identify the caller via ANI or Caller-ID or determine which number was dialed via DID and DNIS. You can also configure your system to discriminate between faxes and ordinary calls.

5.2.2   Placing Outbound Calls

You can specify the following parameters for outbound dialing:

- The number to dial
- The method of call progress analysis to use
- The number of rings and/or the amount of time to wait before timing out

Call Progress Analysis

If you plan to develop an application that makes outbound calls, you should make sure that your hardware provides good call progress analysis, as discussed in Section 4.2.2. After the number is dialed, your application must be able to determine the result of the call. Many professional voice boards have excellent call progress analysis.

Your system should be able to detect the following call progress results:

- No dial tone
- No ringing
- Busy
- No answer
- Operator intercept
- Fax tone
- Connect

Some more sophisticated systems can determine the length of the salutation, indicating who or what picked up the phone: a residence (*"Hello?"*), a corporation (*"Stylus Innovation, may I help you?"*), or an answering machine (*"Hi, this is John. I'm not here right now, so please leave me a message."*).

Call progress signals may vary slightly from phone line to phone line. If your application is having trouble interpreting the result of outbound calls, you should adapt your voice board's call progress analysis. This is common when your computer telephony system interfaces to a PBX. Some voice boards come with software that will analyze your phone system and return a list of definitions for each signal.

Automated Outbound Dialing

If your application places multiple calls in rapid succession, as in a telemarketing or predictive dialing application, you must make sure the phone line is placed on-hook long enough before the next call is started. Otherwise, your phone system will interpret the short period on-hook as a flash hook signal.

Transferring Calls Using a PBX

To transfer calls within your PBX or Centrex service, you need to generate a flash hook signal. This signal puts the phone briefly on-hook, which signals to the telephone system that a service is requested. The flash is generally followed by a string of touch-tones to indicate the extension. For example, the following dial string sent to a Dialogic board will issue a flash, pause, and then dial extension 150:

```
&,150
```

There are two types of transfers: monitored and unmonitored. A monitored transfer makes sure that the other party answers before transferring the call. This requires that your system can perform call progress analysis (see previous section). An unmonitored transfer simply blind dials the extension and then hangs up.

Your system should enable you to change the default lengths for flashes and pauses.

Note: To perform transfers without a PBX or key system, you need additional switching hardware and software, as discussed in Section

5.2.3 Speaking Information

A large portion of any voice application is the speaking of fixed and variable information over the phone. A computer telephony system speaks by playing digitized *voice files* stored on the PC.

Voice Files

A voice file is a recorded message stored on your computer. Generally IVR applications use recorded voice files rather than Text-To-Speech translation because the quality of synthesized voice today is still not as good as a real human voice. (Refer to Section 4.3.2 for more information about Text-To-Speech.)

When playing a fixed user voice file, you can control the following parameters:

- The path and filename of the voice file to play
- Whether touch-tone input can interrupt playback, and which digits will terminate
- The rate and volume of playback
- Where to begin playing in the voice file

Phrases can be saved separately in individual voice files, or saved in a single indexed voice file.

Speaking Variable Information

Recorded voice files can be used to achieve a result similar to Text-To-Speech. Several existing *system voice files* can be strung together to speak variable data, in a technique called *concatenation*. This method can be used to speak commonly spoken items such as numbers and dates. For example, to say the money amount $123.40, your application would need to concatenate the following voice files:

"one hundred" "twenty" "three" "dollars" "and" "forty" "cents"

This allows your application to speak data from a database without needing a voice file for every possible price, part number, or date. For example, your application can speak any number less than one trillion with a vocabulary of fewer than 50 voice files using a few simple algorithms.

You can speak a number in any of the following play methods:

- characters (*"one, two, three"*)
- date and/or time (*"January twenty-third"* or *"Tuesday, 3PM"*)
- currency (*"one dollar and twenty three cents"*)
- number (*"one hundred twenty three"*)
- ordinal (e.g., *"first," "second," "third"*)

In some cases, you might want to specify more information. For numbers and currency, you can specify a precision to which the numbers are spoken (*"One thousand dollars and twenty three cents"* vs. *"One thousand dollars"*). For dates and times, you can specify the format in which the dates are spoken.

To perform concatenation well, you need to trim the beginning and ending of your voice phrases closely to reduce pauses during playback. In addition, your toolkit should play often-used system files from a single indexed file. An indexed voice file contains many phrases one after the other, with an index at its beginning describing where each phrase begins and ends. Using an indexed voice file eliminates the need to perform disk access to open and close each individual voice file, making final playback very smooth and natural without pauses.

Fixed voice files and variable information can be combined to form complete sentences such as *"We have"* [52] *"socket wrenches"* *"in stock, for a total of"* [$104]. These sentences should sound as though they were recorded in one file.

You can either hard-code information into your code or store it in variables or a database. In the previous example, the number of wrenches in stock and their price are stored in the corresponding database record. In addition, the product's name or description (in this case, "socket wrenches") can be recorded in a voice file whose name is indexed in the database table. The application looks up the product description's filename and plays the information at the appropriate time.

Speaking in Multiple Languages

Some computer telephony systems can speak in different languages and voices. You can switch voices at runtime based on a variety of factors:

- Caller input (*e.g.*, menu choice or customer ID number)
- Call identification information (*e.g.*, ANI, Caller-ID, DID, or DNIS)
- Other information (*e.g.*, time of day, keyboard or mouse input)

To speak variable information in different voices, you only need to specify a different set of system phrases recorded in another voice.

In contrast, to speak in different languages, your application must also switch between different sets of concatenation algorithms. Normally, when speaking variable information, your application would use the rules for the American dialect of the English language to determine how to concatenate the system phrases together. The rules for each play method (*e.g.*, currency, date, *etc.*) can be stored in a special language DLL that is accessed at runtime. For example, in French, the number "96" should be spoken as ""four" "score" "sixteen," while the standard American algorithm would speak the number as ""ninety" "six."

Your application should do the following to change languages at runtime:

- Determine the language to use based on caller or other input
- Specify the system voice files used for variable information
- Specify the algorithm for variable information
- Specify a different set of user files for constant information, such as prompts (*e.g.*, by changing the current directory)

In some cases, your application may need to change voices at runtime, but continue to use the default algorithm. Examples include:

- Speaking in a male or female voice
- Speaking in different English accents or dialects
- Speaking in a different language that can use the American English algorithms.

Asynchronous Speech

Generally, a telephony application will play a voice file and then return once the voice file has finished. However, you can play voice files *asynchronously* so that other processing can take place at the same time. This comes in handy when your application is doing heavy processing, such as searching a database, and you want to keep the caller busy in the meantime. For example, your application can place one or more voice files on a queue, return, and then immediately start searching for a database record while the caller hears, "Thank you for your request. Please wait while we look up your records..." Hopefully, soon after your application has completed saying the phrases, it has found the record!

Pausing, Fast-Forwarding, and Rewinding Messages

Each voice file is a certain length, and as the voice file is played, your current position increases towards the end of the file. By specifying or querying the current location within a voice file during recording or playback, your application can pause, fast-forward, or rewind through messages.

To pause, fast-forward, or rewind a message, you must first stop playback and then query the current position in the voice file. To simply pause, set the starting position to the ending position and continue playback. To fast-forward or rewind, set the starting position either larger or smaller than the ending position, respectively, and continue playback.

Note: Some voice boards return the position in bytes, and some return it in seconds. You must convert between the two units based on the number of bytes per minute of recording for the file format you are using. See the next section for a table of voice files formats.

Voice File Storage Requirements

Voice boards can use a variety of compression algorithms to digitize and store voice files. While MCI-compatible sound devices (e.g., SoundBlaster cards) use the common WAV format, most telephony boards (e.g., Dialogic, Rhetorex, NewVoice, BICOM, etc.) use PCM or ADPCM formats (saved with the .VOX extension). Some other boards use a proprietary format, such as IBM Mwave TAM format.

Your application's voice files will generally require from 200K to 500K of disk space per minute of recording, depending on the file format used. Some voice boards let you select between different formats to meet your needs; you must make a tradeoff between disk space and voice quality.

The following table lists common voice file formats and their disk space requirements.

Voice File Format	Quality	Approximate Storage Efficiency
ADPCM 6 KHz	lowest	5.8 min/Mb
ADPCM 8 KHz	:	4.4 min/Mb
PCM 6 KHz	:	2.9 min/Mb
PCM 8 KHz	highest	2.2 min/Mb
IBM TAM compressed	-	3.6 min/Mb
WAV: 8-bit, mono, 11.025*	lowest	2.4 min/Mb
WAV: 16-bit, stereo, 44.100*	highest	0.1 min/Mb

* WAV files can be recorded with a variety of parameters which effect quality and storage requirements, and the above table only lists two combinations. The following table lists the parameters effecting the WAV compression algorithm. (Not all voice and sound boards support all settings.)

Parameter	Valid Settings
Frequency	8000, 11025, 22050, 44100
Bits per sample	8, 16
Mono vs. Stereo	Mono, Stereo

5.2.4   Prompting for Messages

Voice boards can digitize voice and save it to a file. To record a caller's message -- for example, to record a voice mail message or to record special instructions accompanying an order -- you would play a voice file to prompt the caller and then record the caller's voice.

Your application can record multiple voice messages and save a reference to them in a database to be retrieved later. Your application should generate a filename (e.g., using the date and time, incremented integers, or a random number generated by the Visual Basic Rnd() function) before each file is recorded and then save the filename in the database table. You can also determine the date and time of the message (e.g., via the Visual Basic Now() function) and store that with the filename.

The basic parameters for a voice recording are the filename of the message to be stored and the file format (as discussed in the previous section). However, you can configure your system further, as described below.

Terminating a Recording

Your voice board should not continue to record forever; you should configure it so that recording will stop based on various factors. Several ways recording can be terminated include:

- *A terminating touch-tone digit is detected:* You can specify which touch-tone digits will terminate input, if any. For example, you can prompt the caller, *"Please record your name. When you are done, press any key."*

- *The recording times out:* You can set a maximum recording length. You should specify some maximum to make sure that someone does not inadvertantly fill your hard drive with a thirty-minute message!

- *A long pause is detected:* Another indication that a caller is done recording is a long pause. You can specify a maximum length of silence that will be recorded to enable it to guess when the caller is finished speaking.

- *A long constant sound is detected:* Normally human voice is scattered with pauses. A length of solid sound may indicate a machine-generated signal (*e.g.*, a dial tone from the Central Office). To avoid recording such sounds, configure your system to stop recording when a certain length of solid sound is detected.

- *The first burst of sound ends:* You can configure your voice board to stop recording after the first word spoken. This is useful for voice recognition, when your application is trying to recognize single words.

- *The voice board receives a "stop" command:* Your application can stop all voice board activity by sending a stop command.

- *The caller has hung up:* Your system should recognize a line dropped condition and be able to immediately reset itself for the next call.

Other Recording-Related Parameters

The following settings are also useful when recording voice files:

- *Start position:* By setting a start position other than the beginning of the voice file, you can append or partially overwrite existing voice files.

- *Silence compression:* This feature eliminates all pauses from a recording. This is useful for making clean recordings (*e.g.*, when recording system phrases) or for conserving disk space when long pauses are expected (*e.g.*, surveillance applications).

- *Beep:* Whether or not a beep is played before recording begins.

5.2.5 Prompting for Touch-Tone Digits

A computer telephony system can retrieve input from the caller by prompting for digits from the caller's touch-tone keypad. These digits are collected by the voice board and remain in the *digit buffer* until your application retrieves the digits.

Terminating Touch-Tone Input

Your application can use certain criteria to determine how many digits to retrieve. For example, you can specify that the application should retrieve 10 digits; if only 9 digits are in the digits buffer, your system will wait until the 10th digit is received. Whenever digits are retrieved from the buffer, they are also removed. Any digits not retrieved are left in the buffer for later retrieval.

Several ways getting digits can be terminated include:

- *The maximum number of digits is retrieved:* If you expect an exact number of digits, specify a maximum number of digits equal to that number. For example, *"Please enter your ten-digit phone number, area code first."*

• *A terminating touch-tone digit is detected:* You can specify a particular digit that indicates the end of input. So in other words, if you do not know how many digits to expect, you can ask the caller to indicate the end of input with a special key; for example, *"Please enter your order number, followed by the pound sign."* Make sure that the terminating digit is stripped from the input.

• *The retrieving times out:* You can set the maximum time to wait for digits. You should specify some maximum to make sure that the caller has not hung up. Refer to the next section for more information.

• *A long pause is detected:* Another indication that a caller is done entering digits is a long pause between digits. This method is not common but is particularly useful for applications where the touch-tones will be sent rapidly by machine, such as a speed dialer or a barcode-to-touch-tone converter.

• *The voice board receives a "stop" command:* Your application can stop all voice board activity by sending a stop command.

• *The caller has hung up:* Your system should recognize a line dropped condition and be able to immediately reset itself for the next call.

Time Outs

In most situations your application should have time outs set to ensure that the caller responds within a reasonable amount of time and does not tie up the phone line. In some cases, a lack of response may mean that the caller hung up but no disconnect signal was transmitted (see Section 4.2.3 for more details).

You can give the caller a second chance by having the application replay the prompt. For example:

"I'm sorry, I did not hear your response. Please enter your 4-digit PIN code."

After a specified number of time outs, the application should hang up and reset itself.

Other Considerations When Prompting for Digits

You can force the caller to listen through the whole prompt by disabling terminating digits. In most cases, however, you should specify terminating digits that halt prompt playing once the caller starts entering digits. Your system should allow you to specify a different set of terminating digits for retrieving digits and for playing ‛voice files (e.g., the "#" digit terminates getting digits while all digits terminate recording messages).

If your application has multiple levels of menus, as described in the next section, you can specify that digits in the digit buffer are flushed every time a prompt is played. This forces the caller to listen to the beginning of each prompt. To allow more experienced callers to "type ahead" through prompts, configure your system so that digits already in the digit buffer are not lost when the next prompt is played.

5.2.6   Handling Call Flow With Menus

IVR applications commonly use voice *menus* to guide the user through a series of options. A menu allows the caller to choose from a selection of options; for example, *"To hear your account balance, press 1. To make a transfer, press 2. To hear current mortgage and interest rates, press 3..."* The caller presses a key on the keypad to select an option. Then, based on the caller's input, the application will branch to the appropriate set of instructions and execute them.

An easy way to implement menus in Visual Basic is the Select Case statement. The following is an example of a simple main menu for a touch-tone banking application coded with Visual Voice (described in the next chapter):

```
Do While True

        'press 1 if you want to hear your account balance
        'press 2 if you want to transfer to the operator
        'press 3 if you are done.
        Voice1.Value = "prompt.vox"
        Voice1.Action = VV_PLAY_FILE

        Voice1.MaxDigits = 1       'retrieve one digit
        Voice1.Action = VV_GET_DIGITS

        Select Case Voice1.Digits
            Case "1"
                Call Account_Balance ()
```

```
                       Case "2"
                           Call Transfer_Funds ()
                       Case "3"
                           'Thank you for using Bank-at-Home.  Goodbye.
                           Voice1.Value = "bye.vox"
                           Voice1.Action = VV_PLAY_FILE
                           Voice1.Action = VV_HANG_UP
                           Exit Do
                       Case Else
                           '"I'm sorry, that was an invalid selection."
                           Voice1.Value = "invalid.vox"
                           Voice1.Action = VV_PLAY_FILE
                   End Select
             Loop
```

Section 5.3 below discusses important user interface issues to consider when building audio menus.

5.2.7   Client/Server Telephony

To develop client/server telephony applications with Visual Basic, you can either settle for basic messaging and use facilities like the Visual Basic's MAPI control, or, for complete control, add a component to provide network messaging. With a general-purpose network messaging component, you can send messages to client workstations (e.g., when they receive a new call, or when they have a message waiting) and the workstations can send messages back to the telephony server (e.g. to tell the server to ask for the callers name, or to send the caller to voice mail if you do not want the call). Two of the many network communication components include:

- Synergy Software's AppLink VBX
- MicroHelp's Network Lib

For example to use AppLink to send a screen pop message to a workstation named JohnsPC, you would simply add the following lines of code:

```
FaxRcvr.Msg = "You have a new call"
FaxRcvr.MsgLen = Len(FaxRcvr.Msg)
FaxRcvr.Dest = "\\JohnsPC\ScreenApp"
FaxRcvr.Command = SENDMSG
```

To have the application on the client's machine display the screen pop, write 1 line of code in the following AppLink event as follows:

```
Sub FaxSocket_ReceiveMessage (Orig As String, MsgLen As
Long, Msg As String, RetCode As Long)
    MsgBox Msg, "New Message"
End Sub
```

5.3 Telephony User Interface Guidelines

Poorly designed IVR and voice mail applications are infamous for aggravating callers (hence the term "voice mail jail"). You should take special care to follow good judgment when designing your interface.

A whole book could be written about this topic. But as a guideline, keep the following suggestions in mind when you develop your applications:

- *Do not offer too many options per menu.* A good rule of thumb is a maximum of three menu choices.

- *Always give the caller an "escape route."* For example, let the caller press "0" to connect to a live operator or press "*" to immediately return to the main menu.

- *Assign meaning to certain digits and be consistent.* For example, assign "*" to return to the main menu, and "#" to terminate an entry.

- *Explain the option first and the digit to press second.* For example: "For account balances press 1, for transfers press 2..."

- *Allow callers to interrupt voice message playing.* Specify the menu options (or all digits) as terminating digits.

- *Enable experienced callers to "type ahead" through multiple menus.* Configure the voice board so that digits typed before the prompt will not be lost from the digit buffer. (However, do not do this at any ciritcal point that you want to make sure the caller starts listening to the prompt before making a selection.)

- *If there are too many options, allow the caller to spell the choice.* One example is a corporate directory, where callers can determine the extension of the person they are trying to reach by spelling the name. Refer to the voice mail sample application for an example. If you have a limited list of callers, you can instead give them a hardcopy listing of the options (*e.g.*, a list of mutual funds, a catalog of product SKUs, *etc.*).

5.4 Monitoring and Logging Calls

An important part of any telephony application is the accurate monitoring and logging of call information. Examples of information you might want to monitor or log include:

- Number of callers
- Elapsed call time
- Choices selected
- ANI or Caller-ID info
- Documents faxed and received
- Any errors encountered
- Calls abandoned while on hold
- Active lines
- Stage of call (e.g., playing greeting, getting PIN, etc.)
- Number of invalid choices
- Call begin and end times
- To whom the call was transferred

Since this information is very display-intensive, Visual Basic is an excellent environment for quickly putting together a graphical call monitoring window. In such a window, you can display the information in several ways, including:

- *Chronologically in a list box:* A listbox is useful when you need to review the order that events occured or want to watch the system's progress in real-time as calls are processed.

- *Counters in text boxes:* Counters allow you to see at a glance the number of calls, how many times users have selected a particular option, etc.

- *With graphs:* Graphs are useful for viewing varying load levels or for displaying the distribution of callers and selected options quickly and colorfully at a glance. You can easily implement this kind of support with Visual Basic's graph control.

The following is a sample Visual Basic call monitoring application.

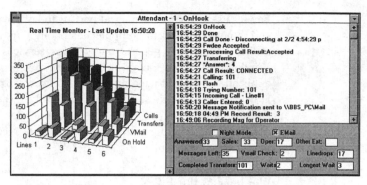

It is wise to also store this information in a database so you can create reports and graphs from the information at any time and view the activity over many days or months. Once again Visual Basic's excellent database connectivity and reporting facilities make it easy to implement this type of data collection and reporting.

5.5 Faxing

The need for fax integration within voice processing applications is rapidly expanding. No business in the 90s is complete without a fax machine, so it is an obvious choice for delivering large amounts of information. You will probably also need faxing to offer your callers some hard copy output, such as a receipt or invoice. You can send or receive existing documents as well as generate documents on-the-fly from another application like a word processor or database.

To add faxing, you can integrate your application with existing fax services, as described below, or use a fax toolkit (refer to Section 6.4.1).

5.5.1 Integration with Fax Services

There are a many different solutions available to perform faxing from Visual Basic. Several vendors sell fax controls which typically require only a few lines to fax out information. For example, with Stylus' Visual Fax controls, you can send a fax using these 3 lines of code.

```
Fax1.FileList(0) = "document.dcx"
Fax1.ToPhoneNumber = "5551212"
Fax1.Action = VF_SEND_FAX
```

87

Faxing has become an integral part of today's work environment and is not restricted to computer telephony. As a result, you can utilize a variety of general purpose fax services (e.g., Microsoft Windows At Work Fax, Delrina's WinFax) from your Windows applications without needing to invest in specialized fax solutions. However, because these services are mostly meant to be used by end users and not by automated applications, you typically sacrifice some programmatic control and receive less status information as a result.

Windows At Work Fax

Microsoft's Windows for Workgroups (WFW) version 3.11 and Windows '95 both include built-in peer-to-peer network faxing called Windows At Work Fax. With this support, you can install a fax modem into any workstation on the network and designate the workstation as a fax server. Other WFW or Windows '95 workstations in the same workgroup can then use the services of the fax server at any time.

Windows At Work Fax is based on MAPI (Messaging API). To send a fax, a Visual Basic application must include the MAPI control, MSMAPI.VBX, which contains two controls: MapiSession to set up a connection to mail, and MapiMessages to read and create individual messages. The Visual Basic application sends a specially addressed e-mail message, Windows At Work Fax will send the corresponding fax.

For example, the following lines of Visual Basic code send the fax message "Hello!" with a subject of "Test Fax" to John Doe at fax # 555-1212. After the server receives the message, it converts it into a fax image and faxes it out to the number specified.

```
MapiSession1.Action = SESSION_SIGNON
MapiMessages1.SessionID = Mapisession1.SessionID
MapiMessages1.Action = MESSAGE_COMPOSE
MapiMessages1.MsgNoteText = "Hello!"
MapiMessages1.MsgSubject = "Test Fax"
MapiMessages1.RecipIndex = 0        'Send fax to n-1 persons
MapiMessages1.RecipDisplayName = "fax:John Doe@555-1212"
MapiMessages1.Action = MESSAGE_RESOLVE_NAME
MapiMessages1.RecipType = 1
MapiMessages1.Action = MESSAGE_SEND
```

Delrina's Win Fax

Delrina's popular WinFax exposes a DDE interface that you can control through Visual Basic. While standard WinFax supports a single fax line, Delrina WinFax Pro for Networks exposes the same interface and supports up to 4 fax lines per server connected on a network. The following Visual Basic code establishes a DDE conversation with WinFax and sends a fax.

```
DDELink.LinkMode = NONE
DDELink.LinkTopic = "Winfax|TRANSMIT"
DDELink.LinkMode = LINK_MANUAL
Call PokeItem(DDELink, "Receiver", "John Doe")
Call PokeItem(DDELink, "Fax Number", "5551212")
Call          PokeItem(DDELink,          "Sendfax",         =
"setcoverpage('smplcvr.cvp')")
Call   PokeItem(DDELink,   "Sendfax",   "fillcoverpage('This
appears on the cover')")
Call PokeItem(DDELink, "Sendfax", "attach('faxfile.fxs')")
DDELink.LinkExecute  "sendfaxui"

Sub PokeItem (DDELink As TextBox, Item$, ByVal Value$)
    DDELink.LinkItem = Item$
    DDELink.Text = Value$
    DDELink.LinkPoke
End Sub
```

5.5.2   Fax-on-Demand

As outlined in Section 5.1, fax applications are as varied as fax broadcasting, fax mail, and fax server. While it is beyond the scope of this book to describe how to develop every one of these applications, this section gives a brief overview of one of the most popular faxing applications, *fax-on-demand*.

A fax-on-demand, or FOD, system allows callers to specify documents to be faxed to them automatically. Generally, a first-time caller asks for a catalog of documents, finds the desired documents in the catalog, calls back, and specifies the documents by touch-tone digits.

A two-call FOD system requests the customer's fax number and then faxes the documents to the fax number specified. A one-call FOD system, on the other hand, sends the fax on the same call. It requires the customer to call from a fax machine, but places the charges for delivering the fax entirely on the customer.

Most fax-on-demand systems have a similar structure. To create your own FOD system:

1. *Configure your faxmodem so that it does not receive incoming calls.* All calls should be answered by the voice board, not the faxmodem. Additionally, the faxmodem should not be tied in by incoming faxes, since it must be available for faxing requested documents out.

2. *When the phone rings, pick up the phone, greet the caller, and prompt for a document number.* If only a few documents are available, your application can prompt with a list of options. Otherwise, you should first inquire whether the caller needs a catalog of documents faxed first.

3. *Translate the document number into a filename.* The document number and filename may be the same, with the exception of the file extension, or you can use a database table to translate the document number into a filename.

4. *Place each filename into an array.*

5. *Loop until the maximum number of documents is reached, or the caller indicates with a touch-tone digit that he or she is finished.*

6. *For two-call fax-on-demand: prompt the caller for the fax number* where the documents are to be sent. Read back the fax number for verification and then ask the caller to hang up.

7. *For one-call fax-on-demand:* prompt the caller to press the Send button on the fax machine to start receiving the fax.

8. *Send the faxes.* For two-call fax-on-demand, hang up and then dial the fax number. For one-call fax-on-demand, simply start sending as soon as the fax tone is detected.

5.6 Handling Multiple Telephone Lines

At some point, you might want to run your telephony applications on more than one line so that your callers do not hear busy signals when the first line is being used. When you develop a multi-line application, you must consider several issues:

- The voice board can return events (e.g., hang up, ring detected, digit pressed, etc) at any time. Your application must respond to these events quickly.

- Your application might process events differently depending on the state of the call. For example, receiving the touch-tone digit "2" can mean "Connect me to sales" at one point in the call, while at another point might mean "Let me leave a message."

- The voice board will return events for many lines, and each line will probably be in a different call state when the event is received.

There are several ways to handle this requirement using Visual Basic, including state machine programming and sequential programming.

5.6.1   State Machine Programming

The state machine development approach has been specially designed to handle the type of requirements particular to multi-line telephony. Unfortunately, it is also one of the most complex and difficult types of programming possible, especially when it comes to reading and maintaining your application logic. As you will see later, easier alternatives exist (see the next section below).

While state machine programming techniques are outside the scope of this book, a quick single-line example may help describe how state machine programming works. Imagine a simple audiotex application that plays and records information based on caller input, with the following call flow:

```
RingDetected ->
Pick Up->
Play Hello ->
GetDigits -> 1 Pressed -> Play Directions ->
             2 Pressed -> Play Company Info ->
             3 Pressed -> Record ->
Hang Up
```

The application would have 10 possible states for each line:

```
RingDetected -> PickUp
PickUp Done-> Play Hello
Play Hello Done -> Get Digits
GetDigits Done & 1 ->    Play Directions
GetDigits Done & 2 ->    Play Company Info
GetDigits Done& 3 -> Record Message
Play Directions Done  -> Hang Up
Play Company Info Done-> Hang Up
```

```
Record Done-> Hang Up
LineDropped -> HangUp
```

The corresponding Visual Basic pseudo-code for this state machine would look like:

```
Sub RingDetected(PhoneLine)
    PlayFile ("Hello", PhoneLine)
    State(PhoneLine) = "Hello"
End Sub

Sub PlayFileDone(PhoneLine)
    Select Case State(PhoneLine)
    Case "Hello"
        Get 1 Digit(PhoneLine)
    Case "Directions", "Company Info"
        Hang Up(PhoneLine)
    End Select
End Sub

Sub GetDigitsDone(PhoneLine)
    Select Case Digit(PhoneLine)
    Case "1"
        PlayFile ("Directions", PhoneLine)
        State(PhoneLine) = "Directions"
    Case "1"
        PlayFile ("Directions", PhoneLine)
    Case "2"
        PlayFile ("Company Info", PhoneLine)
    Case "3"
        RecordFile (Message, PhoneLine)
End Sub

Sub RecordFileDone(PhoneLine)
    Select Case State(PhoneLine)
        Case "Message"
            Hang Up (PhoneLine)
    End Select
End Sub
```

5.6.2   Sequential Programming

If you allow Windows to handle your system's multi-tasking, you can program multi-line applications using ordinary sequential programming. To do this, you need a toolkit that exposes the voice board driver in this manner. Simply develop one or more applications and then run separate instances of each application for each voice board channel. For more information about this type of programming, refer to Chapter 6.

The corresponding sequential Visual Basic pseudo-code for our audiotex application would look like:

```
Sub RingDetected
    PickUp
    PlayFile "Hello"
```

```
        Digits  = GetDigits(1)
        Select Case Digits
            Case "1"
                PlayFile ("Directions")
            Case "2"
                PlayFile("Comapny Info")
            Case "3"
                RecordFile (Message)
        End Select
        HangUp
    End Sub
```

5.7   Testing

While it may seem obvious that you must test any application before moving it into production, this is especially true for telephony applications. Telephony applications are used much differently than others and hence need extra attention during this critical phase. The following table lists some differences between the two:

Typical Application	Telephony Application
Used occasionally	Used constantly
Users are reachable	Users can be anonymous
Users can call a phone number for help	If system goes down, how will they know who to call?
Number of users typically small	A twelve-line system can be called by thousands of users a week
System used during working hours	System used during and after hours, sometimes 24 hours a day
Users can reset themselves after power failures	System must be designed to reset itself after power failures

As illustrated, the typical application is containable and has a smaller impact. In contrast, telephony applications need to be working flawlessly the first time they go live.

So, if it is so important to test these applications, how does one go about doing it? While information on this subject could fill a book on its own, some guidelines are listed below:

Allocate the time to test

Nothing kills a project more quickly than forgetting to adequately test the application. A good rule of thumb is to allocate one third of the time needed to develop an application to testing. When development is nearly complete, readjust the needed testing time appropriately.

Test under heaviest load conditions

If your system can support 24 lines, test with that many. Maybe you'll find that your database does not allow that high a transaction rate and you will have to adjust your system accordingly. As a consequence, you may realize you need a faster machine and you will have to buy and configure it. Unfortunately, you cannot assume the new system will work fine just because you purchased the extra hardware to scale it up!

How can you test under the necessary load conditions? Several approaches vary in accuracy:

- Fake telephony calls with software
- Manually simulate real phone calls
- Automatically simulate phone calls with another telephony system

To fake telephony calls, you can modify your application to start your call flow repeatedly from a loop. Whenever your application requests digits you can feed it random numbers in the range that you need. For example, the following code is a routine to generate different random numbers for different prompts.

```
Function  RandomDigits$  (NumDigits%,  Lower%,  Upper%,
Termdigits$)

    If Lower% <> Upper% Then 'valid range
        Num = Int(Rnd * (Upper% - Lower%)) + Lower%
    ElseIf NumDigits% > 0 Then
        If Termdigits$ <> "" Then
            Num = Int(Rnd * (10 ^ NumDigits%))
        Else
            For I = 1 To NumDigits%
                Number$ = Number$ + Format$(Int(Rnd * 9))
            Next I
            Num = Val(Number$)
        End If
    End If
    Number$ = Format$(Num)
    If TermDigit$ <> "" And Len(Number$) <> NumDigits% Then
        RandomDigits$ = Number$ + TermDigit$
    Else
```

```
            RandomDigits$ = Number$
        End If
    End Function
```

A more accurate approach is to simulate real phone calls into the system. For example, you can ask as many co-workers and friends as possible to try the system and record their problems if any. Similarly, you can arrange a small beta test with a small customer base.

The best approach, however, is to perform realistic automated testing of the environment and have one telephony system call another. To set up such an environment, you can use line simulators attached between voice board channels, or, even better, actually make outbound calls to the real phone numbers that the system will answer. To accomplish this, you can write a specialized application that calls the number, plays touch-tones to your system, and then records the resulting prompts. Your testing application can determine when to fire touch-tone digits if you record a digit at the end of your prompts and have the testing application wait for the touch-tone before responding with the appropriate digits.

You can purchase automated testing tools that will perform these test for you. Refer to Section 8.7 for more information.

CHAPTER SIX

Visual Voice

6.1 Introduction

At this point, you have probably read the previous sections about interfacing to APIs, doing state machine programming, handling concatenation, etc., and are wondering, "Isn't there an easy way to develop telephony applications in Visual Basic?" And the answer is, "Yes, there is!"

While Visual Basic provides a powerful and flexible environment for building applications, you need the right plug-in component to make telephony development easy. Without it, developing telephony applications on your own can be a challenging undertaking. This is one reason our company, Stylus Innovation (Cambridge, MA, (617) 621-9545), introduced Visual Voice in 1993.

Visual Voice is a collection of custom controls and a graphical workbench that make it easy for you to develop powerful telephony applications in Visual Basic. While you can use other methods for developing these applications, Visual Voice makes it easy for you to create telephony applications without learning complex APIs and state machine programming . In a nutshell, Visual Voice provides:

- *A graphical Voice Workbench* allowing you to point-and-click to develop the building blocks of your telephony conversation

- *A familiar custom control interface* for easy sequential development of your application using the Visual Basic language

- *Support for a variety of APIs* including TAPI, Dialogic, and Mwave and *support for advanced telephony features*, including faxing, Text-To-Speech, Voice Recognition, T1 lines, and analog and digital switching.

Developing applications with Visual Voice is straightforward. If you are familiar with Visual Basic, you already know how to use Visual Voice because it works like any other custom control. In addition, you can construct the voice-specific components of your application graphically using the Voice Workbench to.

Developing with Visual Voice is a four-step process:

1 **Visual Voice Custom Control**
Double-click on the Voice control to add it to your project.

2 **Voice Workbench**
Open the Voice Workbench. Point and click to create voice files, strings, queries, and subroutine templates.

3 **Copy and Paste Code**
Paste the generated Visual Basic code into your application.

4 **Add Logic and Run**
Add any additional programming logic to your Visual Basic code. You are ready to run!

If...Then...
Do...Loop...
Select Case...

This chapter gives a very brief overview of how you can use Visual Voice to create telephony applications.

6.2 The Visual Voice Custom Controls

The Visual Voice custom controls provide a high-level interface to voice processing that is consistent with other Visual Basic custom controls. If you are not familiar with Visual Basic programming, the Voice Workbench's point-and-click interface will get you started quickly by generating much of your code for you (see Section 6.3, "Using the Voice Workbench"). The following section discusses how to use the Visual Voice custom controls to add telephony to your applications.

Note: The Visual Voice custom controls conform to the Visual Basic 1.0 custom control specification, which means that they will work with Visual Basic 1.0 and higher. Visual Voice will also work with other tools that support Visual Basic custom controls such as Microsoft's Visual C++, Powersoft's PowerBuilder, Borland C++, and Gupta SQL/Windows. In addition, Visual Voice will soon (at the time of this writing) be available in 16-bit and 32-bit OLE control versions for Windows NT and Windows '95 (refer to Section 2.7).

6.2.1 Adding The Controls

Adding telephony to your Visual Basic application is a simple, three-step process:

1. Add the custom control, VOICE.VBX, to your project. At this point Visual Voice adds several controls to your toolbox.

   If you added the Visual Voice and Visual Fax controls (described later), the toolbox would look like the toolbox below:

Voice Control		Test Control
Link Control		Fax Control
FaxQ List Box		Print Driver Control

2. Add VOICE.BAS, a BASIC file that contains definitions and declarations used by Visual Voice, to your project.

3. Drop the control on your form.

Now you are ready to develop telephony applications in Visual Basic.

6.2.2   Using The Controls Directly

Since Visual Voice is a standard Visual Basic custom control, you can simply set various properties to control its behavior. Also Visual voice notifies your application when telephony events such as when the phone rings via its custom event subroutines. For example, to answer inbound calls you simply place application logic in Visual Voice's RingDetected Event. Any code placed in this event will be executed when the phone rings. Inserting the following line of code picks up the telephone line when the phone rings:

```
Sub Voice1_RingDetected()
    Voice1.Action = VV_PICK_UP
End Sub
```

As you can see, Visual Voice, like other custom controls, allows you to make thing happen by setting an action property to the desired function. In the above example the application picked up the phone line by setting the Action Property to the VV_PICK_UP constant.

There are wide variety of Visual Voice Actions including HangUp, PlayFile, RecordFile, GetDigits, etc. For example, the following code plays a message to the caller and then hangs up:

```
Sub Voice1_RingDetected()
    Voice1.Action = VV_PICK_UP
    Voice1.Value = "Hello.vox"
    Voice1.Action = VV_PLAY_FILE
    Voice1.Action = VV_HANG_UP
End Sub
```

6.3   Using the Voice Workbench

While you can work with the Visual Voice control directly, you can use the Visual Voice Workbench™ side-by-side with Visual Basic to generate corresponding Visual Basic code for all Visual Voice functions. From the Workbench, you can create, edit, and manage four types of *voice objects*:

- Voice files
- Voice queries
- Voice strings
- Voice subroutines

For example, you can use the Voice Workbench to quickly put together a system that speaks the current status of a caller's order from an orders database.

This system:

1. Picks up the phone
2. Greets the caller
3. Prompts the caller for an order number
4. Looks up the record in a database
5. Speaks the order status information (e.g., "shipped yesterday")
6. Hangs up

6.3.1   Creating Voice Files

You can create and record voice files directly from the Voice Workbench.

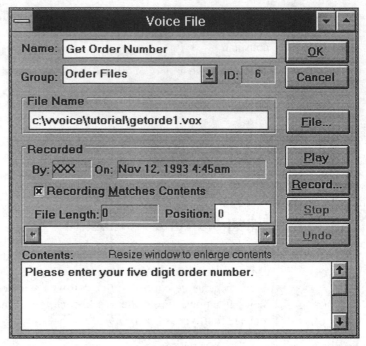

The Workbench keeps track of who recorded the voice file, when it was recorded, the contents of the file, and the voice file length.

You can record or listen to the voice file in this window, or you can record a group of your files at one time using the Batch Recorder. This is especially useful if you have many voice files or if you will use a professional voice-over talent for the recordings. In addition, you can register existing voice files into your project. Also, you can group your voice files according to your own criteria to help keep track of them.

6.3.2   Prompting the Caller with a Voice Query

Voice queries prompt the caller for either touch-tone input or a voice message recording. A voice query consists of two components: the prompt and the expected type of response (e.g., a series of 5-digits, a message, etc.). You can test your voice query directly from this window.

Prompting for Touch-Tone Digits

In the order status system example above, you would prompt the caller for a 5-digit order number. The following voice query illustrates how this is done:

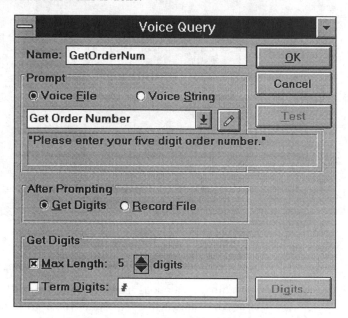

Prompting for a Recording

To instead prompt the caller to leave a message, you would select Record File in the After Prompting radio button group:

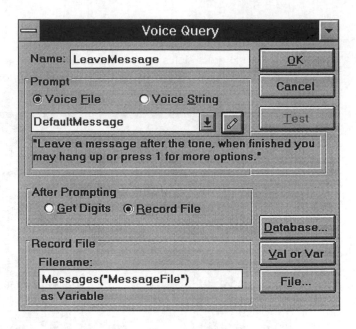

6.3.3 Playing Variable Information with a Voice String

Voice strings are used to combine variable information and voice files into a continuous message or sentence. You can build voice strings by constructing individual voice string elements (e.g., voice files, variables, or constants) and adding them together. For example, to play back the order status, you would build the following voice string: *"Order number" [12345] "was shipped on" [June 7th] "and totaled" [$5.30].*

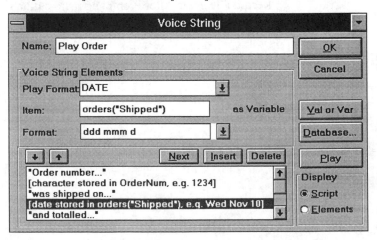

You can click the Play button to test your voice string as you design it. A voice string element, such as the ship date or the total price, can be read from a variable or linked to a database table.

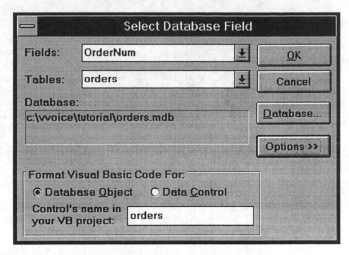

6.3.4   Putting the Pieces Together with a Voice Subroutine

Subroutine templates contain a sequence of actions. In the order status example, your application needs to:

1. Pick up the phone
2. Play a voice file to greet the caller
3. Play a voice query to prompt the caller for the order number
4. Look up the record in the database
5. Play a voice string to speak the order status information
6. Hang up

The following subroutine template initiates each of these actions:

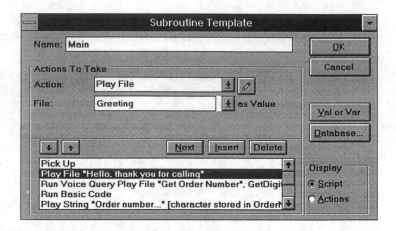

6.3.5   Pasting the Code into Visual Basic

The Voice Workbench automatically generates fully-commented code for you. Simply select Copy Code to Clipboard from the Objects menu, and the code appears on the clipboard.

In this application, you should paste the code into the RingDetected Event to be executed upon any inbound call.

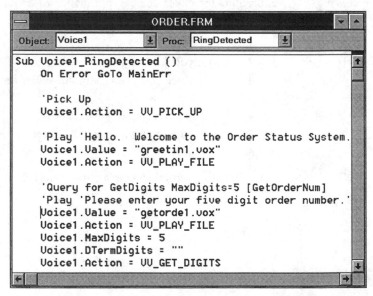

Listed below is all of the code generated by the Voice Workbench that processes the order inquiry. Line numbers have been added next to the code so each line can be explained. Lines without numbers are just Visual Basic comments generated by the Workbench.

```
1.   On Error GoTo MainErr

     'Pick Up
2.   Voice1.Action = VV_PICK_UP

     'Play 'Hello.  Welcome to the Order Status System.'
3.   Voice1.Value = "greetin1.vox"
4.   Voice1.Action = VV_PLAY_FILE

     'Query for GetDigits MaxDigits=5
     'Play 'Please enter your five digit order number.'

5.   Voice1.Value = "getorde1.vox"
6.   Voice1.Action = VV_PLAY_FILE
7.   Voice1.MaxDigits = 5
8.   Voice1.DTermDigits = ""
9.   Voice1.Action = VV_GET_DIGITS
10.  OrderNum = Voice1.Digits

     'Run Basic Code
11.  orders.Seek "=", OrderNum
12.  If orders.NoMatch Then
     'Play 'I'm sorry, that order number is not in the
     orders database'
13.      Voice1.Value = "Isnotva1.vox"
14.      Voice1.Action = VV_PLAY_FILE
15.  Else
     '"Order number..." 1234 "was shipped on..." Wed Nov 10
     "and
     'totalled..." 12.34
16.      Voice1.Value = "OrderNu1.vox|FILE, " & OrderNum &
         "|CHARACTER,wasship1.vox|FILE, " &
     orders("Shipped") &         "|DATE|ddd mmm
     d,andtota1.vox|FILE, " & orders("Total") &
         "|MONEY"
17.  Voice1.Action = VV_PLAY_STRING
18.  End If
     'Play 'Thank you for calling.  Goodbye.'
19.  Voice1.Value = "goodby1.vox"
20.  Voice1.Action = VV_PLAY_FILE

     'Hang Up
21.  Voice1.Action = VV_HANG_UP
22.  Exit Sub

23.  MainErr:
24.  Call vvProcessLineErrors(Voice1)
25.  Exit Sub
```

1   Tell Visual Basic that if there are any errors (e.g., the caller hung up unexpectedly), call a central error handler to hang up the phone and perform any other clean up.

2   Pick up the phone.

3-4   Play the greeting stored in the voice file.

5-6   Play the prompt stored in the voice file.

7-10   Unconditionally get 5 digits, and place them in the Visual Basic variable OrderNum.

11   Search the Orders table for a record with the OrderNum the caller entered.

12-14   If the OrderNum was not found in the table, let the caller know.

16-18   Otherwise, play back the order status via a voice string of files and database variables.

19-22   Say good-bye, hang up, and exit the subroutine to wait for the next call.

23-25   This code will only execute if there is an error. Calling vvProcessLineErrors(), provided with Visual Voice, handles all unexpected errors that could occur.

You can modify this code or add extra logic once you've pasted it into Visual Basic. You can use Visual Basic's extensive debugging facilities to test your application.

You are ready to create an executable. Simply select Make EXE from the File menu, and you are ready to run your application on multiple lines.

Testing with the Virtual Phone

Visual Voice includes a Virtual Phone for testing or demo-ing your applications with any SoundBlaster card. Simply start up the Virtual Phone, and you can call your telephony applications as though you had a real telephone connection. Your application will speak over the speakers and record from the microphone. When your application prompts you, simply click the phone keypad with your mouse or type the digits on your keyboard to enter any touch-tone digits.

6.3.6 Running Your Application On Multiple Lines

If you read Section 5.6.1 and were understandably intimidated at the prospect of state machine programming, take heart! Visual Voice eliminates this worry.

Each telephone line is handled by a separate running copy of the same application. As a result, you can also launch different applications on the same PC. Your application does not have to be aware that other phone calls are being handled. This drastically simplifies your development effort, as you no longer have to track the "state" of multiple phone lines.

You simply write your application for a single telephone line and launch the application on the desired phone lines using Visual Voice's Voice Monitor. Visual Voice can handle up to 24 simultaneous telephone calls per PC.

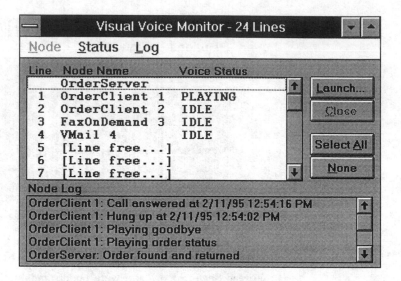

You might think that running several instances of your application would be processor- and memory- intensive, but since it is the same application, Windows only loads one complete copy, and then allocates an additional chuck of memory called a data segment to keep track of the state of each copy. This way, Windows manages the state of each application and line for you, instead of your having to program your application for each state.

6.4   Other Components

Visual Voice works with many types of voice boards and supports a wide variety of advanced telephony features.

6.4.1   Fax Integration

Visual Voice offers several options for fax integration.

Delrina WinFax

As discussed in Section 5.5.1, Delrina's WinFax offers an easy way to interface to most standard fax modems. Visual Voice includes high-level subroutines for sending faxes through WinFax. This solution supports single-line faxing.

Visual Fax

For multi-line faxing applications, you can add Visual Fax, a set of custom controls that add sophisticated faxing capabilities to your Visual Basic applications. You can send, receive, and manage faxes over multiple lines simultaneously, with no polling or IRQs required. A printer driver is provided for faxing from any Windows application. You can monitor fax activity with the listbox control.

Two version are available:

- SatisFAXtion
- GammaLink

Visual Fax for SatisFAXtion works with popular CAS-based SatisFAXtion boards. These boards do all processing on board the fax modem's CPU, freeing up the PC's CPU for other tasks. You can install up to 10 SatisFAXtion boards into one PC. Visual Voice for GammaLink works with single or 4 line GammaFax boards for even higher volume applications with improved error handling.

Mwave Fax

Visual Voice for Mwave includes direct faxing support via an extra fax control. This control interfaces to Mwave's on-board faxing services and offers extra programmatic control over how faxes are sent. For example, you can configure your system to notify your application any time a fax is sent or received.

6.4.2   Sample Applications

Visual Voice includes fully-functional sample applications with source code that you can use as-is or modify for your own needs. Some sample applications include:

- Interactive Voice Response
- Voice mail / auto attendant
- Outbound survey
- One-call and two-call fax-on-demand

In addition, Visual Voice includes a tutorial application that leads you step-by-step through creating your own order status IVR application.

6.5   Visual Voice Versions

Visual Voice comes in three versions to support professional voice boards, multimedia voice/fax/modems, and any TAPI-compatible hardware. 90% of all Visual Voice actions, properties and events are common to all versions, making it easy to create an application that runs on all platforms. However, if you want, you can take advantage of any hardware specific functionality.

6.5.1   Visual Voice Pro

Visual Voice Pro works with Dialogic and Dialogic-compatible voice boards (e.g., Rhetorex, BICOM, New Voice, etc.). Visual Voice Pro supports all advanced telephony features of these voice boards and is the ideal choice for complex multi-line applications. In addition, the superior outbound calling facilities of these boards make Visual Voice Pro the best choice for outbound dialing applications.

Some of the advanced features include:

- Support for 24 lines, including T1 support
- ANI, DID, DNIS, Caller-ID (additional hardware required for certain options)
- Custom tone detection
- Advanced call analysis
- Access to all voice board events and voice board parameters
- International language support
- Voice Recognition
- Text To Speech
- Answering machine and human voice detection for outbound calling applications
- Analog (AMX) and Digital Switch Support (DMX)

6.5.2   Visual Voice for Mwave

Visual Voice for Mwave is primarily targeted towards single users or the small office / home office application.   Visual Voice supports all Mwave-based boards, such as IBM's WindSurfer, Best Data's ACE, and Spectrum's Envoy board.   These boards offer support for 14.4 fax, 14.4 modem, SoundBlaster, MIDI, and telephony at a relatively low price.

6.5.3   Visual Voice for TAPI

Visual Voice for TAPI enables you to create applications for any TAPI-compatible hardware (see the next chapter).   Visual Voice for TAPI is targeted towards the developer who wants to create hardware-independent applications.

Visual Voice for TAPI includes high level commands to complete call transfers and conferences, perform call forwarding, get Caller ID and called numbers, and all other functionality that TAPI exposes.   Although not required, Visual Voice for TAPI includes voice support for any TAPI device that provides a WAV or MCI driver to record and play voice information over the phone line or speakers.

TAPI in Depth

7.1 Introduction to TAPI

The telephony industry has been noted for proprietary interfaces. Hardware and software manufacturers alike have made attempts to standardize telephony development. Recently, *TAPI* has gained significant third-party support. Microsoft and Intel jointly developed the Telephony API (TAPI) to provide a standard interface across hardware platforms. Thus, a TAPI application will be able to run unmodified on any TAPI-compliant board and to work with any TAPI-compliant PBX or switch.

TAPI is actually a piece of a larger Microsoft strategy, WOSA (Windows Open Services Architecture), that is described in Chapter 2. Since WOSA sets common interface standards (e.g., databases, messaging services, financial services, connectivity), TAPI is a natural extension of this strategy.

Microsoft provides a Windows Telephony SDK for developing TAPI-compliant applications. You can use the Telephony SDK with Windows development tools such as C++ and Visual Basic. You can even access a minimal set of telephony functionality using high-level macro and script languages.

To date, (early 1995), a number of TAPI applications and TAPI-compliant hardware platforms have already been released. The number of available TAPI products is expected to expand significantly once Windows '95 is released.

Although you do not have to understand or even use TAPI to develop computer telephony applications in Windows, this book has devoted a chapter to this area because TAPI is emerging as a significant force in the future of Windows telephony.

7.1.1 TAPI Applications

You can use TAPI to create a wide range of applications because it covers a variety of hardware platforms (e.g., modems, PBX's, voice boards, etc.). TAPI lends itself to applications such as:

- *Personal information managers* that offer automatic dialing and voice mail sorting.
- *Groupware applications* through the connection of database managers, spreadsheets, and word processors directly into the telephone network.
- *Software-based phones* that let users manage complicated telephone functions such as coordinating conference calls. For example, to initiate a conference a user could drag three or four names from an electronic directory into a "conference box" and click "connect."
- *Call management* employing Caller-ID to notify users of the identity of an inbound caller via a screen-pop, as described in Section 5.1.5.
- *Remote-control applications* that let users operate their PCs from a distance over public telephone lines.

7.1.2 TAPI Service Providers

Developers use the Telephony Service Provider Interface (TSPI) to build *service providers* for hardware products. For example, voice board manufacturers, PBX vendors, and fax modem manufacturers develop service providers so that TAPI applications can communicate with their hardware. Service providers interpret TAPI functions called by telephony applications and translate them into functions that the underlying hardware can understand.

Microsoft supplies a Telephony DLL component, called TAPI.DLL. TAPI.DLL bridges the Telephony API called by applications and the Telephony SPI implemented by service providers. Thus, TAPI-compliant applications call all TAPI functions directly from TAPI.DLL. In turn, TAPI.DLL routes the instructions to the appropriate service provider for execution. Applications are not required to know anything about the devices that the service providers actually control.

The relationships between TAPI components are shown in the diagram on the following page.

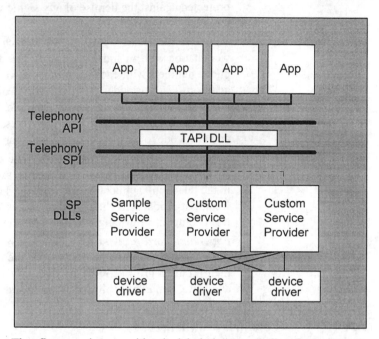

The first service provider is labeled "Sample," referring to the Microsoft-created service provider that ships with TAPI. The sample service provider implements a number of basic telephony functions that work with standard Hayes®-compatible modems. Third-party developers can base their development of more extensive service providers on the sample service provider's code.

7.1.3   Benefits of TAPI

Some of the key benefits of TAPI are listed below. They apply to application developers, hardware vendors, and end users alike.

- *Hardware independence:* As more PCs are equipped with TAPI-compliant hardware, commercial developers can sell their software to run on a variety of installed platforms.

- *Network independence:* TAPI provides support for many types of phone networks, including POTS, ISDN, T1/E1, and PBXs. TAPI applications can transparently run across any supported network.

- *Protection of investment:* End users who invest in applications software and telephony hardware are protected against the demise of any signle vendor by the interchangability of components.

- *Standard addressing:* TAPI provides each supported workstation with an address allowing Windows to match phone numbers with each PC on the network. In addition, TAPI provides "normalization" of phone numbers for outbound calls. For example, TAPI services automatically add the appropriate prefixes, country code, and area code to each outbound phone number.

- *Availability of commercial software:* The appearance of a standard telephony API will lead to increased competition in the telephony application software market. With a wealth of products to choose from, end users will benefit from competitive pricing and feature-rich products.

- *Expanding hardware market:* Telephony network, switch, and hardware vendors will see the markets for their products grow as tools, productivity applications, and vertical market applications are introduced.

7.1.4   Potential Drawbacks of TAPI

Although TAPI offers many benefits, at times it is more appropriate to code directly to a voice board's or PBX's device driver.

The best reasons for you to choose TAPI are hardware independence and the ease of coding to a higher-level TAPI service provider. However, not all voice boards, fax modems, and PBXs support TAPI at this time. Thus, currently the value of hardware independence may be unclear. In addition, TAPI's hardware independence may be of little value to corporate developers that develop and deploy internal applications for a single hardware platform.

A second potential limitation of TAPI is the degree of functionality service providers decide to expose via the Telephony API. In order to claim support for TAPI, vendors need only include support for a limited set of TAPI services, as described later in this chapter. Some vendors may decide to expose only a portion of their hardware's total functionality in their TAPI service providers while providing full access to functionality through their proprietary device drivers.

Essentially, you have to weigh two potentially competing requirements: functionality and hardware independence. If an application requires functionality beyond the basic set of required TAPI services, you should ask your hardware vendor for a complete list of supported TAPI functions before deciding which API to use.

7.1.5 Availability

TAPI is available now for current versions of Windows and Windows for Workgroups. Since TAPI services are not currently installed as part of Windows, they must be distributed with TAPI applications and hardware. Windows '95 will include 32-bit TAPI services out of the box, including small TAPI applications, or applets. Availability of TAPI on Windows NT is expected shortly.

7.2 TAPI Architecture

The Windows Telephony SDK offers a rich array of features, some mandatory and some optional. A system's feature set will depend upon the set of features implemented by both the application and the service provider. The following sections describe the variety of features provided by TAPI.

7.2.1 Telephone Network Services

While the Windows Telephony SDK supports a wide range of telephone network services (e.g., POTS, ISDN (Integrated Services Digital Network), T1/E1, Switched 56, Centrex™, PBXs, and key systems), TAPI enables applications to handle all of them transparently. Applications use only the abstraction of "lines" and "phones" without needing to consider the underlying network.

119

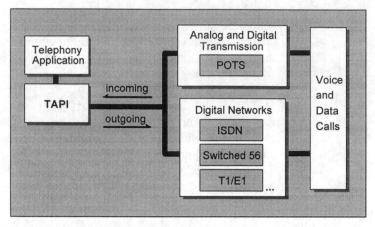

Because TAPI is independent of the underlying telephone network, programming a PBX application using TAPI is the same as programming a POTS application using TAPI. You can use an application originally programmed for a POTS environment within a PBX environment with no changes to the application's source code. However, to take advantage of additional features offered by networks such as PBXs and ISDN, the task of developing TAPI applications would become more complex.

7.2.2   Device Classes

In the TAPI programming model, different real-world objects such as telephones, modems, and telephone lines are considered *device classes* at the API level. Thus, you can treat each member of a given device class in a similar manner. The class framework is extensible, allowing TAPI to classify and support equipment developed in the future.

The Windows Telephony API currently defines two device classes: the *line device class* and the *phone device class*. It also defines two sets of functions and messages, one used for line devices and one used for phone devices.

Physically speaking, a line device, such as a fax board, a modem, or an ISDN card, is connected to an actual telephone line. Line devices allow applications to send or receive information to or from a telephone network. A line device contains a set of one or more channels or ports that can be used to establish calls. Within TAPI applications, a line device is the logical representation of a physical line device; TAPI views a line device as a point of entry that leads to the switch.

Just as a line device class is an abstraction of a physical line device, the phone device class represents a device-independent abstraction of a telephone set.

7.2.3   TAPI Call Control

Telephone lines and phone sets are connected to each other logically through TAPI. The physical connection can be made at the desktop, or through a LAN-based host or server, where a LAN protocol "extends" the connection of the phone lines or phone to the client application. The following diagram illustrates a physical connection made at the desktop.

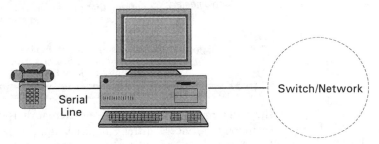

Applications access TAPI services using *first-party* call control. In other words, the application controls telephone calls as if it were an "endpoint" (the initiator or the recipient) of the call. The application can make calls, be notified about inbound calls, answer inbound calls, invoke switch features such as hold, transfer, conference, pickup, and park, and can detect and generate touch-tones for signaling remote equipment.

In contrast, a controlling application using *third-party* call control would not act as an endpoint of the call. Third-party call-control allows an application to establish or answer a call between any two parties, but the application does not act as either of these parties.

7.2.4   The Functional Hierarchy of Windows Telephony

Windows Telephony services are divided into *Assisted Telephony* services and the services provided by the full Telephony API. In general, you would use the full Telephony API to implement powerful telephony applications, while you can utilize the simpler Assisted Telephony services to add minimal dialing functionality to non-telephony applications.

Windows Telephony's services are divided into the groups shown below.

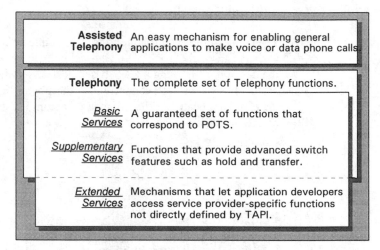

| Assisted Telephony | An easy mechanism for enabling general applications to make voice or data phone calls. |

| Telephony | The complete set of Telephony functions. |

Basic Services	A guaranteed set of functions that correspond to POTS.
Supplementary Services	Functions that provide advanced switch features such as hold and transfer.
Extended Services	Mechanisms that let application developers access service provider-specific functions not directly defined by TAPI.

The programming models of these different divisions of Windows Telephony services differ significantly. Because Assisted Telephony and the full Telephony API are used and implemented in different ways, mixing Assisted Telephony function calls and TAPI function calls within a single application is not recommended.

Assisted Telephony

Assisted Telephony enables the establishing of voice and media calls from any Windows application, not just from those applications dedicated to telephony. Assisted Telephony lets applications make telephone calls without needing to work with the full Telephony API. This small set of functions extends telephony to word processors, spreadsheets, databases, personal information managers, and other non-telephony applications.

For example, adding the Assisted Telephony function **tapiRequestMakeCall** to a spreadsheet lets users automatically dial telephone numbers stored in the spreadsheet or in a connected database. As long as the application needs none of the detailed call control provided by the full Telephony API, Assisted Telephony is the easiest and most efficient way to give it telephonic functionality. Capabilities beyond dialing, such as the transmission and reception of data, require additional data transfer APIs, including the communications functions of the Comm API.

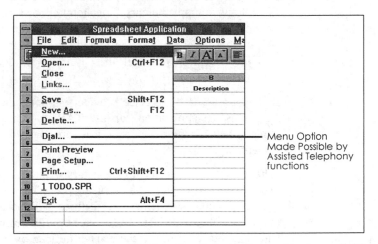

Assisted Telephony adds four functions to those already available to the developer in other APIs such as the Windows SDK.

The Telephony API: Three Service Levels

Applications that offer more than the most basic outbound call control require the Telephony API. The Telephony API interface consists of all Windows Telephony's functions *except* those of Assisted Telephony. In reality, the full Telephony API contains all the functionality of Assisted Telephony, but the commands are different. TAPI defines three levels of service:

- Basic Telephony
- Supplementary Telephony
- Extended Telephony

Basic Telephony Services

Basic Services are a minimal subset of the Windows Telephony specification. Since all service providers must support the functions of Basic Services, applications that use only these functions will work with any TAPI-compliant device. Basic Telephony functionality includes line handling, placing calls, answering calls, etc.

Supplementary Telephony Services

Supplementary Telephony includes all services defined by the Telephony API beyond those included in the required set of Basic Telephony. Supplementary Telephony exposes features found on modern PBXs, such as hold, transfer, conference, park, and so on. All of these features are considered optional; that is, the service provider decides which of these services it does or does not provide.

For example, if the service provider supports the supplementary functions, an application can use TAPI to transfer and forward calls. All phone-device services are part of Supplementary Telephony.

The behavior of each supplementary feature is defined by the Telephony API, not the service provider developer. This guarantees that supplementary features will behave as expected across telephony devices. If a service provider cannot implement a supplementary service in the manner defined by TAPI, the feature should instead be provided as an extended telephony service.

Extended Telephony Services

Service provider vendors can enlist *Extended Services* to extend the Telephony API with their own device-specific extensions. Extended Services (or Device-Specific Services) include any extensions to the API defined by a particular service provider. For example, PBX vendors can expose features specific to their equipment via TAPI extended services. Special functions and messages such as **lineDevSpecific** and **phoneDevSpecific** are provided in the API to allow an application to directly communicate with a service provider. Since the API defines the extension mechanism only, the definition of the Extended-Service behavior must be completely specified and documented by the service provider.

7.2.5   Extending TAPI with Media-Stream APIs

TAPI by itself provides control only for line and phone devices and does not give access to information exchanged over a call. To manage the media stream itself, you need to integrate functions from other APIs into your telephony applications. Examples of Windows-based media control APIs include:

- Multimedia wave audio (WAV)
- Media Control Interface (MCI)
- Data APIs

For example, an application that provides an interface for managing fax or data transmission would use TAPI functions to control and monitor the line over which bits are sent, but the actual data would be transmitted with functions from another API, such as the communications functions (known collectively as the Comm API) of the Windows SDK. (See the Windows SDK documentation for more information about controlling data communication devices.)

In the same manner, the media stream in a speech call is produced and controlled by one human talking to another. (While voice consists of any signal that can travel over a 3.1 kHz-bandwidth channel, speech refers exclusively to human speech.) The WAV API can play and record voice files and consequently handle part of the voice stream transmitted over a TAPI-controlled line. In any case, the line over which the call is established and monitored, and the call itself, remain in control of the TAPI application.

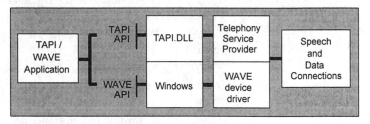

Hardware Considerations

In order to effectively run an application, the underlying hardware must not only support TAPI, but any other media control APIs involved. For example, a simple voice mail application would require multiple functionality from the underlying voice board or PBX:

- Support for Basic and Supplementary Telephony services, including answering calls, detecting digits for menu navigation, terminating the call, etc.
- MCI or WAV drivers for playing and recording voice messages

If a voice board or PBX does not include some type of sound driver, you lose the ability to manage speech media. Likewise, in order for a TAPI application to transfer data, the underlying hardware must support both TAPI services and communications functions.

7.3 Programming TAPI Applications

To use the Windows Telephony SDK, you will need the Windows SDK, including a C compiler, linker, resource editors, resource compiler, and debugging and optimization tools. You can also access the Telephony API directly from Visual Basic. Currently, only the development of 16-bit applications is supported.

7.3.1 Elements of the Telephony SDK

The Telephony SDK is available from Microsoft on CompuServe, free of charge. The SDK includes the following elements:

- *TAPI.DLL:* A dynamic link library that routes TAPI functions between applications and service providers.
- *TAPI.LIB:* A library that provides entry points to the functions of TAPI.DLL.
- *TAPIEXE.EXE:* An application that is automatically loaded by TAPI.DLL and provides an application context for handling telephony events.
- *The Dialer:* A sample call-manager application that provides the most basic telephonic functionality, such as processing requests from applications to make voice calls. (This processing ability makes it an Assisted Telephony server.) It also has its own user interface that lets users make calls directly. Full source code for the Dialer is provided, and this source code is the starting point from

which most programmers will begin programming their own telephony applications.

. *Sample Service Provider (ATSP.C):* Sample source code that can establish and control a simple call using a Hayes ®-compatible modem. Its main purpose is to help test Windows telephony applications and to provide sample source code for service-provider writers.

. *The Telephony Control Panel (TELEPHON.CPL):* A control panel applet that is used through the Windows Control Panel to install, remove, and configure telephony service providers and to set location and calling-card information used in address translation.

. A *number-translation module (TAPIADDR.DLL):* A dynamic link library that translates addresses in to "dial-able" format.

. *Header Files* that include TAPI.H for API functions and TSPI.H for SPI functions.

. *Electronic Documentation* that includes the *Telephony Application Programmer's Guide* and the *Telephony Service Provider Programmer's Guide.* This documentation is provided electronically in Microsoft Word for Windows version 2.0 format, and can be printed on almost any laser printer. No hardcopy documentation is provided with the Windows Telephony SDK.

. *Help files* for both application programmers and service provider programmers. These Windows Help files contain complete on-line function references of all Windows Telephony API and SPI functions.

7.3.2   Visual Basic and TAPI

You can access the entire Telephony API from Visual Basic. Visual Basic is most suited to creating telephony applications, not service providers. To access TAPI directly from Visual Basic, you must add a header file to your program to declare all TAPI variables and subroutines. Developers can download a Visual Basic header file from CompuServe. (See Chapter 6 for an alternative method of developing TAPI applications with Visual Basic.)

Like all Windows applications, telephony applications operate in the event-driven model used by Windows. In TAPI applications, the most important events are call-state transitions, which can be divided into two categories: solicited and unsolicited. A *solicited* event is one caused by the application controlling the call (i.e., when it invokes TAPI operations), while *unsolicited* events are caused by the switch, the telephone network, the user pressing buttons on the local phone, or the actions of the remote party.

Visual Basic does not have direct access to Windows events (callbacks) on its own. However, products like SpyWorks-VB from Desaware can expose these Windows messages to Visual Basic programs.

Visual Basic Example

The sample code below demonstrates adding Assisted Telephony functionality to a Visual Basic application to enable it to make an outbound call. For a complete explanation of Assisted Telephony see the Section 7.2.4 earlier in this chapter.

The simplest Assisted Telephony operation is **tapiRequestMakeCall**. This function enables an application to pass a telephone number and related information to the Dialer application included with the Telephony SDK.

The example below shows how this function could be used in a simple Visual Basic application. The sample application accepts a phone number and an optional name string (separated by a slash) from the DOS command line. Program Manager icons could run this application to dial frequently-used numbers. You would simply place the appropriate parameters in the Command Line field of the Preferences dialog associated with the icon.

First, you must declare the function and any constants used. This is done in the general declarations section of the form:

```
Declare Function tapiRequestMakeCall& Lib "TAPI.DLL" (ByVal
DestAddress$, ByVal AppName$, ByVal CalledParty$, ByVal
Comment$)
Global Const TAPIERR_NOREQUESTRECIPIENT = -2&
Global Const TAPIERR_REQUESTQUEUEFULL = -3&
Global Const TAPIERR_INVALDESTADDRESS = -4&
```

The application fetches the contents of the command line used to launch the program:

```
Sub Form_Load ()
    a$ = Command$
```

Next, it searches for a "/" (which would indicate a name to go in the Call Log follows), and if one is present, parses the command line into the address to dial (a$) and the name (n$):

```
n$ = ""
i = InStr(a$, "/")
If i <> 0 Then
    n$ = Trim$(Mid$(a$, i + 1))
    a$ = Trim$(Left$(a$, i - 1))
End If
```

It then passes these fields to tapiRequestMakeCall. If the result is 0, that means that the request has been accepted; it is up to the Dialer application to do any further work.

```
retval& = tapiRequestMakeCall&(a$, "", n$, "")
If retval& = 0& Then End
```

If an error occurs (non-zero result), you can display a dialog box to explain what happened.)

7.3.3   TAPI: Looking Forward

The Microsoft/Intel Telephony API holds great promise for application developers who require hardware portability. For TAPI to be truly successful, hardware vendors such as PBX and voice board manufacturers will need to provide TAPI interfaces for their products that rival the features offered in their proprietary interfaces. Numerous influential companies, including AT&T, Connectware, IBM, Mitel, Northern Telecom, OCTuS, Siemens, Sierra Semiconductor, Stylus Innovation, and Toshiba, are enthusiastically adopting TAPI as a key strategic technology and are developing products that embrace the TAPI standard.

Microsoft and Intel are committed to the success of their Telephony API. Microsoft will distribute TAPI support with upcoming 32-bit Windows operating systems. Microsoft is supporting TAPI developers by providing a free and easily accessible SDK that you can access from C, as well as from Visual Basic. Microsoft and Intel also sponsor multiple inter-operability events ("Bake-offs"), where telephony application and service provider developers can meet and test their respective products prior to release. The turn-out of hardware and software developers at the last two bake-offs points to a burgeoning market for TAPI products in 1995 and 1996.

CHAPTER EIGHT

Telephony Hardware

8.1   Introduction

As mentioned earlier, most PCs or other computers are capable of running IVR applications as long as they are equipped with voice processing hardware. This chapter provides an introduction to the many hardware options that are available today but by all means does not cover every vendor or hardware component in the industry.

8.2   Voice Processing Boards

The following sections list vendors for both professional voice boards and multimedia voice/fax/modems. For more information on selecting a voice board, refer to Section 4.4.2.

8.2.1   Professional Voice Board Vendors

Today, the majority of computer telephony systems take advantage of the advanced features and reliability of professional multi-line voice boards. Handling from two to thirty ports each, several multi-line boards can be combined within a total system to accommodate even more telephone lines. Professional voice boards vary in supported features.

The following is a partial list of professional voice board manufacturers:

BICOM
Trumbull, CT
203.268.4484

Pika
Kanata, Ont,
Canada
(613) 591-1555

**Dialogic
Corporation**
Parsippany, NJ
1.800.755.4444 or
201.993.3030

Rhetorex
Campbell, CA
408.370.0881

New Voice
Vienna, VA
703.448.0570

8.2.2   Multimedia Voice/Fax/Modem Vendors

Voice/fax/modems primarily target the single-line user and SOHO
markets. They tend to be relatively inexpensive boards with a
great deal of functionality bundled in one package. Most
voice/fax/modems are built upon one of three technologies:

- IBM's Mwave®
- AT&T®'s VCOS®
- AT+V

IBM's Mwave

Today, the Mwave series of single-line voice boards provide a
powerful multimedia environment for creating sophisticated voice
and fax applications. This software-upgradeable Mwave board
includes support for voice, fax, data modem, SoundBlaster®, wave
table, MIDI, and Caller-ID capabilities. Mwave support for two
lines is currently under development, and according to IBM,
should be available sometime 1995.

The following is a partial list of Mwave-based hardware
manufacturers:

IBM Corporation
800.IBM.2YOU
WindSurfer board

**Best Data
Products, Inc.**
818.773.9600
*Advanced
Communication
Enhancement
(ACE) board*

Spectrum
604.421.5422
Envoy board
Solo board

ObJIX
Microsystems
Corporation
617-466-8720
Media Manager

AT&T's VCOS

Developed by AT&T, the Visible Caching Operating System, or VCOS, is at the center of an emerging group of DSP-based multimedia boards. Currently, VCOS boards that can handle voice/fax/data tasks are only available in single-line versions. However, the VCOS technology allows multiple boards to be stacked together allowing multi-line applications. Another nice feature of the VCOS telephony support is WAV files can be played and recorded over the directly over the phone line.

The following is a partial list of VCOS-based hardware manufacturers:

American
Megatrends
Norcross, GA
404-263-8181
Mediatel board

IPC Corporation
Ltd.
Singapore
265 744 2688
Telemetry 32 board

Communications
Automation and
Control (CAC)
Allentown, PA
18103
610-776-6669
QUANTUM DSP
board

AT+V Voice/Fax/Modems

Several chip manufacturers have created a voice extension to the Hayes AT command set called AT+V. Just as Class 1 and Class 2.0 specifications added fax support to the AT command set, the AT+V specification adds voice recording and playback as well as touch-tone decoding to the AT string-based command set. Using this specification, the board transfers voice files over the serial port, which currently translates to fairly poor quality voice because of the limited bandwidth serial ports provide. Future versions of this specification aim to improve voice quality and add additional telephony features. Because these boards are not DSP based, they are extremely inexpensive and popular, with list prices below $100. The flip-side is that they are not currently software upgradeable.

The following is a partial list of AT+V voice/fax/modem manufacturers:

Boca
Boca Raton, FL
(407) 997-6227
Multimedia Voice
Modem

Zoom Telephonics
Boston, MA
(617) 423-1072

8.3 Fax Modem Manufacturers

The following is a partial list of fax modem vendors:

BICOM
Trumbull, CT
203.268.4484

Brooktrout
Technology
Needham, MA
(617) 449-4100

Dialogic
Corporation
Parsippany, NJ
1.800.755.4444 or
201.993.3030
GammaLink Boards

Pure Data
Richmond Hill,
Ontario
(905) 731-6444
Intel SatisFAXtion

Rhetorex
Campbell, CA
408.370.0881

8.4 TAPI Service Providers

As discussed in detail in Section 7.1.2, TAPI-compliant hardware must have a service provider that interfaces the hardware driver to TAPI-compliant software. Although TAPI is an emerging standard, there are already many companies developing service providers. Some service providers are developed by the hardware manufacturer, while others have been developed by third parties.

Some sources for services providers are listed below, along with the type of hardware they support:

Answersoft
Plano, TX
multimedia
voice/fax/modem

AT&T Bell Labs
Middletown, NJ
PBX

AT&T Global
Business Solns
Middletown, NJ
PBX/Phonesets

AT&T VCOS
RTP, NC
multimedia
voice/fax/modem

Comdial Enterprise
Systems
Charlottesville, NC
(804) 978-2406
multimedia
voice/fax/modem

Connectware
Richardson, TX
multimedia
voice/fax/modem

Ericsson Business
Networks
Tyreso, Sweden
PBX

Genesys
San Bruna, CA
PBX

Harris
Melbourne, FL
PBX

Intecom
Dallas, TX
PBX

Inter-Tel
Chandler, AZ
PBX

ISDN Systems
Vienna, VA
PBX

Mediatrends
Concord, MA
multimedia
voice/fax/modems

Microsoft
Redmond, WA
AT modem SPI

Mitel Corp.
Kanata, Ontario,
Canada
PBX/ phonesets

Northern Telecom
Minnetonka, MN
PBX and others

OXUS Research
Meudon La Foret
Cedex, France
ISDN PC cards

Panasonic
PBX

Rhetorex
Campbell, CA
voice board

Rolm
Santa Clara, CA
PBX

Sound Designs
Israel
*external voice
hardware*

**Spectrum Signal
Processing**
Burnaby, BC
*multimedia
voice/fax/modem*

Toshiba
Irvine, CA
PBX

**Voice Technologies
Group**
Buffalo, NY
PBX card

Xinex
New Westminster,
BC, Canada
phone

8.5   Switching Solutions

Switching boards provide the ability for sharing devices among several voice lines. They allow applications to provide transferring and conferencing capabilities without a PBX or share a single fax board between multiple voice lines. For further information on switching, see Section 4.3.1. The following is a partial list of switching hardware manufacturers:

Amtelco
McFarland, WI
608.838.4194
XDS Switch

**Dialogic
Corporation**
Parsippany, NJ
1.800.755.4444 or
201.993.3030
*AMX, DMX, MSI
boards*

Rhetorex
Campbell, CA
408.370.0881

Dianatel
San Jose, CA
408.428.1000
SmartSwitch boards

8.6   Other Hardware

8.6.1   Text-To-Speech Solutions

As mentioned in Section 4.3.2, Text-To-Speech functionality can be achieved using three methods: software, firmware, and hardware.

TTS Software

Software-only products load Text-To-Speech algorithms into the PC's CPU, which performs all of the processing. The following lists some of the available software vendors:

Berkeley Speech Technologies, Inc.
BeSTspeech
Berkeley, CA
510.841.5083

Lernout & Hauspie
Woburn, MA
617.932.4118

First Byte
ProVoice

TTS Firmware

Firmware products can be loaded onto a voice board's processor, reducing the load on the PC's CPU. The following vendors provide TTS firmware:

Berkeley Speech Technologies, Inc.
BeSTspeech
Berkeley, CA
510.841.5083

TTS Hardware

Text-To-Speech hardware is an additional board that shares the voice lines via a switching board or the voice bus. The following lists some of the available TTS boards vendors:

Digital Equipment Corporation
Merrimack, NH
DECTalk

Infovox
Sweden

8.6.2 Voice Recognition Solutions

Voice recognition enables your system to understand words spoken by a caller. For more details, see Section 4.3.3. VR is useful for reaching phones that do not have touch-tone service. Some of the available solutions are as follows:

Dialogic
Corporation
Parsippany, NJ
1.800.755.4444 or
201.993.3030

Lernout &
Hauspie
Woburn, MA
617.932.4118

Scott Instruments
Corporation
Dallas, TX

817.387.9514

Telaccount Inc.
Bronx, NY
718.824.3493

Voice Control
Systems
Dallas, TX
214.386.0300

Voice Processing
Corporation
Cambridge, MA
617.494.0100

8.6.3 Pulse-to-Tone Converters

Most voice processing hardware provide support for only touch-tone services. For those that need to access rotary or pulse services, many pulse-to-tone converters are available.

Aerotel
Holon, Israel
972.3.559.3222

PTC-30 Series: internal boards supporting 4 or 8 channels

BICOM
Trumbull, CT
(203) 268-4484
Aerotel daughter card

PIKA Technologies, Inc.
Kanata, Ontario
(613) 591-1555
PTX Series: external box supporting 2, 4, 8, or 16 channels

Rhetorex
Campbell, CA
(408) 370- 0881
RDSP voice board: includes pulse recognition

Hager International
Natick, MA
(508) 651-8310
Sumihiro PT02 and PT04z

TeleLiaison
St. Laurent, Quebec
514.333.5333
Genius Series: external box supporting 4 or 8 channels

8.6.4   Call Identification Solutions

Call identification services (ANI, Caller-ID, DNIS, DID) is supported by a variety of hardware and software solutions. Some voice boards already include some call identification features, while other features require additional hardware. For more details on call identification, see Section 4.3.4. The following is a list of vendors that offer call identification solutions.

Comdial
Enterprise
Systems
Charlottesville, NC
(804) 978-2406
Caller-ID on voice board

DAC Systems
Shelton, CT
(203) 924-7000
Caller-ID

Dialogic
Corporation
Parsippany, NJ
1.800.755.4444 or
201.993.3030
DNIS, ANI

EXACOM Inc.
603.228.0706
Concord, NH
Caller-ID, DID,
DNIS, and more

Rochelle
Communications
Austin, TX
512.339.8188
Caller-ID

Vive Synergies
Richmond Hill, Ont
(905) 882-6107
Caller-ID

Mountain Systems
Grundy, VA
(703) 935-2275
Caller-ID

8.6.5   Line Simulators

The following vendors provide a variety of line simulator
equipment.

Alltel Supply, Inc.
Norcross, GA
(404) 448-5210
Viking Products

Dialogic
Corporation
Parsippany, NJ
(800) 755-4444 or
(201) 993-3030
PromptMaster™

Northeast
Innovations
Concord, NH
(603) 736-8260
Series 1040

Teltone
Bothell, WA
(206) 487-1515
COLS (w/Caller-
ID)

8.7   Testing Systems

For more information about automated testing systems for telephony applications, contact:

Hammer Technologies
Wilmington, MA
(508) 694-9959
The Hammer

Index